Mating Males

Examining mating from the male perspective, this book provides an overview of mammalian reproduction to illustrate the important role that males with a desire to mate play in the life of mammals.

Written in a conversational style that will appeal to those without specialist knowledge of the field, it begins with a broad overview of sexual reproduction in mammals, explaining the importance of mixing genes, sexual selection and the concept of mating seasons. Subsequent chapters examine some of the most important aspects in detail, including mating behaviour, the structure and function of the male organs of reproduction and their physiological control and modes of copulation. It also challenges some conventionally held notions about testicular descent and scrotal function and presents some new thoughts and ideas on these subjects.

A final chapter considers human reproduction, explaining how our physical and social evolution have contributed to the development of sexual behaviour that is markedly different from that of other mammals, due in particular to the absence of oestrus and seasonality in the human female.

TIM GLOVER is Emeritus Professor of Veterinary Anatomy at the University of Queensland. He has written over 100 papers on male reproduction and is the co-editor of *Male Fertility and Infertility* (Cambridge University Press, 1999).

Hyperion – by Gainsborough out of Selene. One of the most famous thoroughbred sires of all time (if not arguably the most famous). He was a chestnut and stood only 15 hands high (1 hand is just over 11 cm).

Mating Males

An Evolutionary Perspective on Mammalian Reproduction

TIMOTHY GLOVER

University of Queensland, Australia

CAMBRIDGE
UNIVERSITY PRESS

Shaftesbury Road, Cambridge CB2 8EA, United Kingdom

One Liberty Plaza, 20th Floor, New York, NY 10006, USA

477 Williamstown Road, Port Melbourne, VIC 3207, Australia

314–321, 3rd Floor, Plot 3, Splendor Forum, Jasola District Centre, New Delhi – 110025, India

103 Penang Road, #05–06/07, Visioncrest Commercial, Singapore 238467

Cambridge University Press is part of Cambridge University Press & Assessment, a department of the University of Cambridge.

We share the University's mission to contribute to society through the pursuit of education, learning and research at the highest international levels of excellence.

www.cambridge.org
Information on this title: www.cambridge.org/9780521159579

© Timothy Glover 2012

First published 2012

A catalogue record for this publication is available from the British Library

Library of Congress Cataloging-in-Publication data
Glover, Timothy D.
  Mating males : an evolutionary perspective on mammalian reproduction / by Timothy Glover.
    p. cm.
  ISBN 978-1-107-00001-8 (Hardback) – ISBN 978-0-521-15957-9 (Paperback)
  1. Mammals–Reproduction. 2. Mammals–Sexual behavior–Evolution.
3. Generative organs, Male–Evolution. 4. Men–Sexual behavior.
5. Sexual behavior in animals. I. Title.
  QP251.G524 2011
  599.156–dc23

                                                          2011018533

ISBN   978-1-107-00001-8   Hardback
ISBN   978-0-521-15957-9   Paperback

To Claire

Contents

Preface

In some quarters it is fashionable to try and argue that mating males (especially human ones) are surplus to requirements when it comes to the progress and continuation of their species. That animals can be cloned and that technology enables some men without sperms in their ejaculate to father children appear to confirm this. But this book aims to dispel any such contention and to explain how, through mating, a male mammal fulfils an indispensable role in the maintenance of its own species.

Reproduction by means of cloning can only have a stultifying effect in the long run, and is ultimately likely to be lethal. Like those of a clockwork toy, the mechanisms of survival need to be wound up from time to time, if for no other reason than to survive the vagaries of the environment. In mammals, including man, this is achieved by the periodic injection of new and randomly different genes through sexual reproduction. The evolutionary development of sex has thereby introduced an extra dimension to the natural selection (survival of the fittest through accidental mutation) of asexually reproducing species.

In mammals, the unison of eggs and sperms, each having half of the normal complement of chromosomes, has facilitated a reduction in the incidence of too many morbid mutations, whilst enhancing the possibility of good ones being introduced. The importance of this arrangement is that favourable characteristics can be achieved in one generation, or a few, without significantly affecting the phenotype (overall body characteristics) of the species. Survival in mammals has depended upon this method of ensuring adequate genetic variation. It is facilitated by having males in the population in addition to females. The wider the mix of genes by different types within a species, or even from a closely related species, the tougher the offspring are likely to be. Cross breeding (*hybridization*) between different species of baboons has demonstrated that new and doubtless beneficial characteristics can soon make their appearance in a group.

Farmers have known about the importance of this sexual genetic variation for years and have continuously cross bred sheep, for example, because they have been aware of the principle of so-called *hybrid vigour*. This means that cross breeds are often sturdier than inbred animals. Crossing Swaledale ewes with a ram such as a Dorset Horn, for instance, yields what are called 'mules', sheep which withstand the harshness of the Yorkshire winters best, whilst producing meat and wool of the highest quality. There are other examples, exemplified, for instance, by improving the hunting capacity or augmenting scent skills in dogs. Also, it is particularly noteworthy that mongrels are much more likely to enjoy better health than closely inbred animals.

By contrast, it is well known that inbreeding, and even worse, cloning, is not desirable, if the aim in the long term is to produce really good healthy stock. Reducing or eliminating genetic variation, therefore, has not proved to be biologically advantageous, and in the wild there are usually plenty of unrelated males wandering around to ensure that it doesn't happen. Rarely, for example, does one see male cheetah siblings mating with the same female. If they did so too often, it would be contrary to the principle of genetic variation. Why distribute most of your genes with a particular female if your brother has already done it for you? For the sake of genetic variation, it is much better to go off and mate with an entirely different female in order to increase genetic mix. A single cloning in cows, sheep or any other animal can be very useful, but probably not far beyond one generation. The production of 'Dolly' was a fantastic achievement. However, I think it unwise to over-interpret the potential of cloning and of producing transgenic animals, even though our ability to do so has been an amazing advance in biological knowledge.

Victorians recognized that it could be inadvisable for close relatives to marry. It was commonly held that if cousins married, they would be likely to yield offspring who were mentally deficient or lunatic. This is not necessarily so, of course, but was among the mores of the time, presumably intended to discourage too much inbreeding. It was already understood all that time ago, that to reproduce your kind with an initial stranger is more likely to improve the stock – hence our distaste for incest. An apparently unrecognized flaw in the philosophy of 'planned eugenics', which was quite fashionable among some intellectuals before the Second World War, was that reduction in the randomness of the gene mix can only end in tears. People in Sardinia appear, at first sight, to be an exception, in that they are long lived and yet go in for quite a bit of inbreeding. But human longevity is a complex business and not a good example of natural survival.

The most striking example of the adverse effects of inbreeding is to be seen among pedigree dogs. By breeding dogs from parental and first-generation animals or from siblings, abnormalities of conformation and even genetic disease are perpetuated or created. Fashions at Crufts (not entirely their fault) have turned many excellent basic breeds into cripples. Governments are unlikely to legislate against this, lest the self-righteous indignation of those blinkered breeders and their societies leads to electoral disadvantage.

But what is the point or the sense of legislating against docking spaniels' and boxers' tails, whilst allowing Pekingese to be bred so they cannot breathe properly? To breed dogs that are so specialized that they cannot mate spontaneously must be the ultimate in absurdity. If something is not done, it must be obvious that many good breeds of dog will die out altogether, with an awful lot of suffering en route. Readers might gather that I feel strongly about this! Micro-pigs also pander to human whim, yet society is up in arms at the very mention of designer babies. What's good for the goose is surely good for the gander!

In addition to providing for more immediate survival of a species in new and alien environments, the mixing of genes with attendant mutations can be augmented and added to naturally, if more randomly, over very long periods and thus, in typical Darwinian fashion, create a new species. But new species cannot naturally be created overnight and may indeed take thousands of years. Natural speciation is generally a very slow process.

The means whereby genetic amphimixis in mammals is assured (through spontaneous mating) is discussed in the pages that follow.

This book is about mammals – reference to other animals is only made out of passing interest or for comparison. This is simply because my main interest over the past 50 years or more has been in the biology of mammals and I feel ill-qualified to write about other animals in any detail. This is not to imply that they are any less fascinating. I have also confined my attentions to the male, because most of my earlier work was focused on sperm production and the interpretation of semen quality. In this I was privileged at Cambridge to have three great masters of reproductive physiology as my mentors. These were Sir John Hammond, Dr Arthur Walton (pioneers of artificial insemination in British cattle) and Professor Thaddeus Mann. Professor Mann was an unswervingly supportive personal friend for many years after Sir John and Dr Walton died. I am deeply indebted to all of them and hope they would have enjoyed this presentation (though doubtless, they would have had their criticisms!).

In this text I examine the structure and arrangement of the organs of reproduction, not just for the sake of it, but rather as a demonstration of how their shape and form in different species of mammals reflect reproductive strategies and sexual tactics that have evolved over the ages according to biological circumstance. The final aim of every sexually active male is to reproduce, but details of the means of doing so differ in different species, even though mating is always involved. Only the intellectually deprived, those people who are unable to recognize overwhelming scientific evidence when they see it, can fail to accept and embrace the principles of evolution through natural selection, as set out by Darwin in his work on the origin of species. So, in examining differences in structure and form we need to seek evidence for how it might have happened.

I hope that by approaching the subject of mating in different mammalian species, I might be able to offer a new slant on the way we look at the process in ourselves. A word of caution is needed, however. Whilst this book is not for those who giggle at the very mention of mating, nor for those who consider the subject vulgar and not to be mentioned at all, it is certainly not for the prurient. Neither, though, is it for the squeamish or faint-hearted. This is not a drawing-room story, but is intended to be a glance at what is a really interesting evolutionary development in the spectrum of biological functions.

The book is a light canter through the field and, in the final chapter, I expressed some personal opinions. For several years I worked as an andrologist in an infertility clinic and I have to make clear that some of these opinions are based on clinical experience and anecdotal evidence rather than experimental data. I have also served as a professional witness in a variety of court cases and some of my remarks in the last chapter are drawn from that experience. So, although this chapter may in parts appear to be unduly empirical, its intention is to provide ideas and some new ways of looking at the subject of human reproduction. Apart from this, some of the issues raised and conclusions reached, whilst based on scientific evidence and observation, may be regarded by some authorities as being debatable. But ideas and discussion of them must surely be one route to getting things right, and differences in interpretation can be useful in this context. The text is written deliberately by way of a discourse, in the hope that it might be more readable than a more conventional work and, if nothing else, might periodically entertain, rather than induce somnolence. The book, you see, is not really intended for bedtime reading.

I have not interrupted the narrative by inserting hundreds of references into the text, more scholarly though that might appear to have been. This is not only because I have not wanted to spoil the flow of the story, but also because much of the material is based on papers written in the last century. By and large, it seems to me, contemporary biologists rarely read papers written more than about ten years back, so listing older papers might be regarded as a waste of time. But I have given just a brief list of key works at the end of each chapter for anyone, especially students, who might wish to take things further. In so doing, I am putting in a plea for older works to be consulted frequently, because a broad background can only provide better perspective.

It has not been too easy to choose the most appropriate and useful references, because the book consists of a distillation of information acquired over many years of observation and reading, and much of it is well known. Many, if not most, of the references cited are to books or reviews, and if a student dips into these, many more references to individual papers will be found. A few of these, including some of my own, are provided when I have considered them to be of particular relevance in understanding the subject under discussion. The list does not, therefore, necessarily refer to specific points made in the text, but is rather provided for wider general reading. I hope the text might prove to be of sufficient intrinsic interest to those outside the field, without the need for further reading, so addressing the references is simply a matter of choice or need.

Domesticated and laboratory animals receive most attention in this book. This is because, in the context of the subjects discussed, more is known about them than other mammals. It could also reflect a certain bias, due to my having been reared as a boy on a farm, and having bred rabbits and rats in my schooldays. This is not to mention my frequently being asked, on our farm, to hold mares while the stallion covered (served) them. Much of the information presented here on wild mammals stems from my own observations at zoos, especially Chester Zoo in the United Kingdom, and in the African countryside. This was Africa when you could spend a whole day in the bush (or 'bundu') and never see a human being, but could alone share the rather savage beauty of the Rift Valley with the indigenous animals. In those days, there were no Volkswagens painted like zebras or chattering tourists with their frantically clicking cameras to mar the curious majesty of Kenya, a quite magical place.

Writing objectively about mating is not always easy without appearing indelicate. I think this is largely because a lot of people interpret animal behaviour anthropopathically – they ascribe the feelings and passions of man to other animals. I have a clear memory of a photographic exhibition in Liverpool in the United Kingdom, where I once worked, showing two butterflies in sexual union. The caption read 'Butterflies fornicating'. Maybe Liverpool butterflies do fornicate, but it seemed rather anthropopathic to me!

Occasionally, I have toyed with a little anthropopathy in this book, but only humorously or as an explanatory metaphor. If we want a broad and meaningful perspective on the behaviour of non-primate mammals, it is better not to try and connect it with human behaviour at all. Worse still, we tend to think it is funny to be anthropomorphic and put animal heads on humans or dress animals up to make them appear like humans. Most of us have indulged a little in this sort of thing, and Walt Disney did it rather charmingly and respectfully. We also see it in circuses. It is a sort of habit, and it existed even before Aesop's fables. For example, sexual activity for tasteful human beings is a private affair, and in writing this book I have occasionally felt guilty, lest I might have intruded too far into the private lives of other mammals. But we can be assured that no such sense of privacy is likely to prevail among those mammals when a female is in heat.

Research into DNA relationships between mammalian species is very useful in all sorts of ways, but that we share some patterns of DNA with pigs and guinea pigs does not mean to say that we are all that close to them or that we behave like them. We even share aspects of our DNA with some invertebrates. I honestly believe, and this is not a condescending comment at all, that scientists must be cautious not to mislead the public in efforts to communicate their science. Without details, it is easy for people to get hold of the wrong end of the stick. It should no longer be necessary to produce further evidence of Darwinism (no longer a mere theory). Of course, we are likely to have some aspects of our DNA which relate to that of other animals, but that is to be expected during evolution. We are descended from a sub-human primate, but this does not mean that we need to behave like other sub-human primates. It is puzzling why this is so difficult for some people to grasp.

Yet magazines and television programmes, in particular, are repeatedly tempted into trying to relate the sexual behaviour of wild animals to that of man, and vice versa. This is wrong and very misleading. Indeed, the final chapter of this book explains how human mating behaviour is likely to be unique among mammals as a result of the special nature of human evolution. A major theme in this book, for instance, is that animals that have no periods of sexual desire (periods of heat or oestrus) behave quite differently from those that have.

In writing this material, I have tried my best not to offend, but if there are parts of the text where I have failed, I hope any reader will understand the difficulty and be forgiving. In trying to arrive at a suitable balance, a certain moral detachment might show through. But the book is about biology, and it is intended to be informative, yet thought provoking.

Timothy Glover
Brisbane, Queensland, 2011

Acknowledgements

First, I want to thank all those good scientists who have contributed to knowledge in the field of reproductive biology. I have usually found their work, whether presented verbally or in print, to be arresting and absorbing. I apologize at the outset to the authors of many splendid works that have not been cited in the listed references of this book. The constraints of space and logistics have been pressing and, therefore, I decided that I would simply try to provide an entrée to each subject covered in the text. Thereby, the gate can be opened into the field and a deeper look at the literature should then provide a pathway across it. Information on this subject is extensive and experts will recognize that such a compact volume as this can only be a broad pastiche. Of course, I want to thank those whose books and reviews have been cited here. I believe they will prove a very valuable guide to those who wish to look into the subject more deeply.

However, there are some people whose special help leads me to name them. This I do gladly. I am greatly indebted to Ron Hunter of Edinburgh for his relentless encouragement and support. Without this, and, therefore, without him, I doubt whether the book would ever have been completed. I am appreciative of the help received from Keith Harrison and Frank Carrick of Brisbane. I am also particularly grateful to 'Twink' Allen, former Professor of Equine Reproduction at Cambridge for permitting me to use the drawing shown in Figure 5.2, and members of his staff for their help and interest and for permitting me to use the photographs shown in Figures 2.5 and 5.7, and to Roger Short and Jim Cummins for their interest and help. I am indebted also to Sandra Wilsher for the

photograph of the statue of Hyperion, which appears as a frontispiece
and for her great help with other figures.

It is a friend indeed, who can be bothered to read one's material
and so I want to express my sincere appreciation to Michael Bryden,
Emeritus Professor of Veterinary Anatomy at the University of
Sydney, and John Tesh of the University of Surrey, UK, for agreeing
to help me in this way for helpful suggestions.

I wish to thank Joy Schonrock of the Biological Sciences Library
in the University of Queensland for her painstaking assistance with
references, Howard Munro and my son Ben Glover for assistance
with what to me are the mysteries of computerization, John Morgan
for his unfailing encouragement and for permitting me to use
equipment in St John's College in Queensland and Maureen Crisp
for help with photocopying. My thanks are also due to Lulu Skidmore
of the Camel Reproduction Unit, Dubai, UAE, for permission to
reproduce the photograph of camels mating (Figure 5.15), Lee Morris
of New Zealand for permission to use her photograph (Figures 2.7
and 5.1) and, in particular, Ian Robinson for his tireless and
indispensible help with proofs and the index.

Most of the anatomical drawings are the work of Doug Bailey
when he worked in my laboratory at the University of Queensland,
and I am most appreciative of being able to use them. This has been
possible through the courtesy of Paul Mills of the University of
Queensland, who is currently Director of Teaching and Learning
in the Veterinary School.

It has also been a privilege to have so many interesting
discussions over the years with the staff at Bellevue Zoo, near
Chester, UK, and to have enjoyed the friendliness and encouragement
given to me in past years by the zoo's director, the late George
Mottershead.

Finally, I thank Cambridge University Press for publishing this
material for me and for their patient assistance.

1   The system

Since this book is about mammals, it is reasonable to ask the question: 'What is a mammal?'. A mammal is a vertebrate, which means it has a vertebral column or backbone – a facetious answer to the question could be that a mammal is a mammal when it is not a fish, an amphibian, a reptile or a bird. All of these are also vertebrates, you see. Most mammals have hair, wool or fur, but not all; some have spines and others scales, whilst man has little or no such covering. So there is great variety among mammals, which makes it all the more confusing when it comes to definition. However, the correct answer to the question is that mammals are those creatures that suckle their young. This distinguishes them from other animals, including the other vertebrates mentioned.

Reproductively speaking, there are three kinds of mammal. There are those that lay eggs, such as the duck-billed platypus and echidna (what used to be called the spiny anteater), those that give birth to embryos – the marsupials, including kangaroos among others (you can accommodate up to a dozen newly born opossums on a teaspoon) – and those such as ourselves, who give birth to relatively well-developed foetuses after nurturing them for longer periods in the womb (the uterus). Each of these groups is classified sequentially as being *prototherian*, *metatherian* or *eutherian* mammals. Thus, the completion of body form occurs in the uterus of eutherian mammals in contrast to that of a 'joey' (baby kangaroo or any other marsupial), for example, whose longest period of development is at the teat, having been born as a young embryo.

Nevertheless, there is quite a lot of variation in the degree of development of newly born eutherian mammals. Guinea pigs, or cavies as they are sometimes known, are virtually independent as soon

FIGURE 1.1 A rather curious mammal! The strange and mysterious platypus lays eggs, but although she has no teats, she suckles her young. Her milk is ejected over the surface of her abdomen for the young ones to drink. © Photolibrary.com

as they are born, whereas the newly born of other species – such as humans, elephants or pandas – need a much longer period of maternal care before they are capable of fending for themselves. Animals that give birth to live young whose earliest development takes place in the uterus are known as being *viviparous* and typify eutherian mammals, while mammals that lay eggs (i.e. the monotremes, such as the duckbilled platypus – Figure 1.1) are, like birds, referred to as being *oviparous*. Some fishes and reptiles (e.g. the adder) are viviparous; while some fishes are said to be *ovoviviparous* (or oviviviparous), because they provide advanced development in the egg and yet give birth to live young.

The above information is all by way of background explanation, but let's get back to mating. It is often equated with the act of copulation, which means 'to couple' or 'be fastened together'. But mating may have wider meaning than this; sniffing around or foreplay is also part of the process and, in some species, mutual care of young between male and female may come into it. So perhaps mating is best regarded as a drawing together of male and female for purposes of successful sexual reproduction.

By mating with different females, a male is able to scatter his genes widely, and since most mammals mate with several females rather than one, the dispersal of his genes may be quite extensive. His aim is to outdo his peers in the distribution of his own genes. From a purely biological standpoint, a male mammal may be thought of as simply being testicles on legs, because its body functions mainly to transport the testicles from place to place. This function is all-important in sexual reproduction, since it is the testicles that produce the gene-bearing sperms.

The more females a male impregnates, the greater, we say, is his reproductive fitness, because thereby, the wider his genetic stamp will be. Moreover, such a successful male will provide a greater mix of genes (whilst sustaining his personal hallmark in the offspring), than his less successful rival, who may mate with fewer females or none at all.

It is important to recognize that reproduction involving two individuals is by no means confined to vertebrates, let alone to mammals. The transfer of nuclear material from one individual to another is even to be seen in single-celled organisms such as the paramecium and single strands of the simple plant, spyrogyra (conjugation). Furthermore, RNA and DNA are not confined to nuclei, so we can have bacteria and viruses, which are without a nucleus, spontaneously reproducing and sending modification messages between them. But actual sexual reproduction (the union of eggs and sperms) occurs commonly among invertebrate multicellular organisms, including worms, insects and other arthropods, so beetles, moths and butterflies, as well as a host of marine animals including octopuses and their relatives, and crabs, lobsters and their relatives, are included. In some animal species, sexual reproduction appears to have been tried and then discarded during their evolutionary history. In others, such as the malarial parasite, it is periodically introduced into what is essentially an asexual reproductive cycle. But sexual reproduction seems to have been fairly efficient, because it is so widespread throughout the animal kingdom, even, for example, in hermaphrodite worms, such as the tapeworm.

With the exception of hermaphrodites, the popularity, as it were, of sexual means of propagation is most likely to be due to the flexibility and versatility that results from a mostly random mixing of genes. It has been argued that this notion needs a rethink, but until we get one, the principle makes reasonable sense. This genetic amphimixis, however, demands that sperms and eggs (in mammals, each carrying half the number of chromosomes carried by other cells in the body) unite in order to yield viable offspring. The success of the system depends on their being some biological assurance that this will happen.

PROTECTION OF SPERMS AND EGGS

Male and female sex cells (gametes) cannot unite if they are unable to survive long enough to do so. This is obvious, but it does mean that they need to avoid environments that are fiercely adverse to their wellbeing. In addition, it is clearly desirable that they should not be deposited too far from each other. If these needs can be met, then their chances of meeting each other in good condition are increased and they can get away with a relatively short lifespan. Thus, we see that sperms and eggs are not only protected in terms of space and territory, but also by timing and individual longevity.

Spatial protection is achieved by males and females moving closer together as the time for sperm and egg deposition approaches. As far as temporal protection is concerned, we find that reproductive activity tends to be intermittent rather than constant, and in a large number of animals it is confined to certain seasons of the year. This seasonal activity has a side-effect, in that it also saves on the number of sperms and eggs necessary for the process to work. This can be important, because producing eggs, in particular, requires considerable energy.

Most males produce a prodigious number of sperms, which is a powerful instrument in ensuring that, in spite of inevitable wastage, there remains enough of them to reach the site of fertilization. Nevertheless, there are safety mechanisms built into the system, because the life and survival of a sperm can be precarious. For example, the sperms of some species can resist differing environmental conditions. But even so, when sperms and eggs are deposited in a watery environment – as they are in most fishes, where fertilization is external to the body – dilution by the surrounding medium, predation and purely adventitious wastage are all hazards that sperms and eggs must face. Extremes of temperature, serious changes in pH (acidity and alkalinity) or osmolarity are not likely to be easily tolerated, so the environment needs to be stable enough to be congenial to these reproductive cells. The sperms of many marine animals are isosmotic with sea water (that is, they do not draw in or push out more fluid than is good for them), but fishes that live in either fresh water or sea water might find one more conducive to the protection of their gametes than the other. Salmon, for example, move from sea water to fresh water for spawning. Thus, there is a level of vulnerability beyond which gametes will not be able to survive, so the adult animals that produce them must behave accordingly.

Sexual rituals, by both some vertebrate and invertebrate males, lure females to come as close to potentially fertilizing sperms as possible. In amphibians such as salamanders and axolotls, the male literally guides his female to sperms that he has deposited on the ground. Eventually she will snatch them up into her genital opening or cloaca. In the case of the salamander, she gets her directions to the sperms through a male dance of courtship. Newts also induce their females to pick up sperm packages in claspers that are situated in their hind parts. Among the invertebrates, male dragonflies do an alluring dance and scorpions have a similar, if slightly more complicated ritual; among vertebrates, female salmon emit a chemical that blends with another one from the male and attracts him to come closer to her. Spatial protection of sperms and eggs, therefore, can sometimes be a combined effort by both sexes.

In most of these animals, and we see it in other vertebrate and invertebrate species, as well as in some cartilaginous fishes, there is an added protection for sperms in that they are encased, as if wrapped up in a parcel, before being called upon to go forth and fertilize. These wrapping cases are called 'spermatophores'.

The invertebrate octopus, several species of shark (all species of which have a backbone consisting entirely of cartilage rather than bone) and other fish, including many bony fishes, produce spermatophores. There are some marine invertebrates, however, which agglutinate their sperms during internal transportation, so that groups of them stick together in small packages without being wrapped up. They are only released to be individually free when they hit the sea water.

At first sight, it might seem that when fertilization takes place within the female body (so-called internal fertilization), with sperms being deposited directly into the female tract, the sperms are especially well protected and that this is the safest procedure of all for them. However, it turns out that this is more apparent than real, because the female tract is not always as hospitable to invading sperms as might be imagined. On the contrary, provided the sperms are shed fairly close to eggs, external fertilization in water is probably safer. It may come as something of a surprise to know that there is at least one species of fish that releases 1000 times fewer sperms than are to be found in the ejaculate of a reindeer, for example. Surely, this discrepancy in numbers suggests that there are more hidden dangers for sperms in the female tract of a mammal than there may be for those released into the sea. With internal fertilization, though, there must be less risk of sperms wandering off in the wrong direction and getting lost.

In mammals, there are particular regions of the female tract that are less of a menace to sperms than others. Indeed, some areas of the tract may be regarded as safe havens. In some invertebrates such as crabs, there are receptacles in the female tract that provide such havens, and in mammals the neck of the womb or the cervix and the junction between the uterus and the Fallopian tubes (uterotubal junction) are less dangerous regions for sperms than other parts of the tract such as the vagina or the body or horns of the uterus itself. As might be expected, mammalian sperms move out of the trickiest environments within the female tract as soon as they can and head for these safer places. There are exceptions, as in biology there nearly always are! Some bats store sperms in the wall of the uterus, a normally hostile place, for up to six months, but this is an unusual specialization for mammals.

When fertilization is internal, sperms can be deposited into the female in a variety of ways. They may be injected under the skin as if from a syringe, or limbs may be used for insertion, as in the hectocotyledonous arm of the octopus. Insemination in most birds and some fishes is achieved by apposition of the male and female cloacae, although some birds have a phallus or penis – the ostrich and the duck are the most notable, but there are a few others. These penes have a ventral sulcus or groove, rather like the embryonic phallus of a mammal, so they do not actually ejaculate through it, though semen may in part be directed by it. The penis of the waterfowl is a very elaborate affair, designed, it seems, to try and overcome structural barriers within the female. This underscores the fact that birds, as we shall see in the next chapter, can be quite selective in the males they permit to impregnate them. So, in sexual reproduction, the existence of a penis is not consistent in all species and it seems first to have evolved in some of the flatworms (platyhelminths).

SYNCHRONIZING MALES AND FEMALES
IN MAMMALIAN SPECIES

Internal fertilization and its attendant need for copulation characterizes mammals, so when the time is ripe, males and females need to draw closer together, with a view to being very close. This is because their form of mating involves bodily contact between the sexes. To achieve this, the male must draw attention to himself and encourage the female to accept such a physical liaison. In the majority of mammals this acceptance is only periodic and is restricted to her periods of desire – when she is 'in heat' or 'oestrus'. These periods might only occur in a mating season and are non-existent out of season. Some mammals have discarded the idea of mating in specific seasons of the year, but to argue that all mammals are fundamentally seasonal breeders, or have tended to be so in the past, is compelling.

In sexual reproduction it is obviously essential that the release of sperms and eggs is properly synchronized, and this means that copulation should occur at a particular and appropriate time. In mammals, this synchrony is achieved by the ovary releasing eggs during or shortly after the periods of a female's acceptance of a male. The release constitutes the process of **ovulation**. It ensures that a female receives sperms at a time which corresponds closely to the release of her eggs. But the length and frequency of the female's period of desire and, therefore, the opportunity for copulation, vary widely between mammalian species.

Horses, dogs and cats (and their relatives in the wild) have a fairly long period of oestrus, whilst in cows and ewes, it is only short. Cows may be in heat for only five or six hours in the winter, and it might not even be noticed, but in the summer it usually lasts for about a day. Oestrus in ewes in the northern hemisphere is confined to a breeding season between October or November and about March (except for the Dorset Horn in Britain, which breeds twice per year) and it lasts between one and two days only. Sows have a slightly longer period of two to three days.

However, within the mating period, be it seasonal, intermittent or constant throughout the year, the frequency with which females come into heat or oestrus also differs between different species. The period between one heat and another is known as an 'oestrous cycle'. In horses and cats, this is about three weeks long, but ewes come into heat every 16 days. There are quite a lot of heat periods in these animals and they are thus referred to as being 'polyoestrous'. In the last century, the Hunters' Improvement Society (HIS) registered stallions used to travel around the country by road, covering mares at different stables and farms. They would call back in three weeks and repeat the procedure if the mares had again come into oestrus. This was rather unreliable, since mares can ovulate (and thus come into heat) in early pregnancy or even with foal at foot, but the method seemed to work quite successfully.

Dogs – and their wild relatives such as wolves and foxes – only come into heat every year or every six months, making them

FIGURE 1.2 'Mighty Power' by Radium out of The Waif (a descendant of the great St Simon, through his dam). Champion HIS thoroughbred premium stallion in the 1930s (see text). He was also a chestnut, but stood 16.5 hands high.

'monoestrous' animals, although their carnivorous relatives, the hyenas, are polyoestrous. Among the smaller laboratory animals, rats have a distinct oestrous cycle, but rabbits do not. If rabbits are not receptive to a buck, they are always either pregnant or displaying false pregnancy (which is quite a common condition in rabbits). In most other mammals, though, there is a regular oestrous cycle and this results from an equally regular cycle of activity of the ovary (Table 1.1).

The majority of mammals ovulate spontaneously during their periods of oestrus (cows being an exception in that they ovulate shortly after their heat period has ended), but a number of mammals are unable to do so. Ovulation in these species has to be induced and this is normally achieved by the male. In these cases, the female cannot ovulate unless coitus or copulation occurs. Included among

Table 1.1 *Oestrous periods and cycles*

Species	Length of oestrous cycle (days)	Length of oestrous period (days)
Mare	21 (polyoestrous)	7
Cow	21 (polyoestrous)	1 or <1
Sow	21 (polyoestrous)	2–3
Ewe	16 (polyoestrous)	1.5
Goat (Nanny)	16 (polyoestrous)	1
Bitch (Dog)	180–240 (monoestrous) (prolonged period of anoestrus – see text)	7–8
Queen (Cat)	21 (polyoestrous)	2–3

these 'induced' or 'non-spontaneous' ovulators are all the cats, including the lynx and jaguar, together with better-known wild cats such as the tiger, leopard and lion, as well as feral and domestic cats. Many smaller mammals such as rabbits, hares, minks, ferrets, raccoons, squirrels, field mice, voles and shrews are also induced ovulators, as are some larger mammals such as camels (including also llamas and alpacas). The females of these species need a copulating male to fire off the ovulatory hormone (luteinizing hormone or LH) from the pituitary gland at the base of the brain, and thus to bring about the release of their eggs. It is a very good way of making sure that the release of eggs and sperms is synchronized, even though egg release often does not occur for several hours after copulation has taken place.

Among the wild cats, lions copulate about every 15 minutes throughout a day. As will be seen later, it is possible that each mount, and apparent male orgasm, might not yield sperms. If this is the case, we cannot so far tell whether the first or first few mounts end in the ejaculation of sperms or whether it is the last or last few that do so. But there can be little doubt that copulation in induced ovulators is not solely concerned with the delivery of semen; the repeated coitus of lions might bring about a sort of staircase stimulus to the pituitary gland of the females.

FIGURE 1.3 A male lion of Africa. If competitors have been seen off and he has no need for food, this mighty animal's attention will be focused on the lioness, provided she and/or her cohorts are in heat.

In a number of mammals, there is some bleeding from the female tract before the onset of oestrus. It is seen in some of the primates, such as the tree shrew, but most notably in bitches. This is not to be confused with menstruation. The bleeding during this pro-oestrous phase of the cycle (that which occurs immediately before the oestrus phase itself) is a simple rupture of uterine blood vessels, whereas menstruation involves a major breakdown of uterine tissue. Neither is associated with the period of female desire, although it has been noted that in some monkeys, copulation takes place more frequently during the menstrual period. In the pro-oestrous phase, bitches are often very skittish in the presence of a male, but will usually sit down very firmly if he tries to mount her at this time.

Most of the New World monkeys (those in Central and South America) do not menstruate, in contrast to menstruation being typical of the Old World monkeys of Africa and Asia. These include macaque monkeys among others, such as gibbons and other great apes. Menstruation is also, of course, a human characteristic. It is important to recognize, therefore, that a menstrual cycle is not the same as an oestrous cycle.

The significance of menstruation is not clear, although I have some of my own ideas about it. Perhaps most of our female forebears were always pregnant, in which case menstruation might be considered as being an abnormality, a pathological condition, if you like. But this is a separate story and it is mating that is under discussion here.

SEASONAL MATING

From what we have discussed so far, it is clear that the female is the partner who decides how often mating will take place, and she is the one who restricts it. A female, after all, puts far more energy into the reproductive process than her male counterpart, so it is inadvisable to waste too much of it on mating. Eggs, for instance, are much larger than sperms and since they must first divide after being fertilized without any serious direct maternal help, they need some food of their own in the form of yolk. Eggs of birds and reptiles have a great deal of it, but mammalian eggs have relatively little and our own have virtually none at all. However much yolk is produced or not produced, cell division requires an energy input and thus, when fertilization has occurred, it is the sex cells within the female (pre-embryos) that need it. This has to come originally from the mother. By contrast, the sperm as such disappears after fertilization, and in any case their production is doubtless less energy-consuming than the manufacture of eggs in the first place.

Pregnancy in itself is also energy intensive. Some mammalian species choose to expend most of their energy on carrying several young per gestation (so-called 'polytocous' mammals), whilst others carry singletons or twins for a longer period ('monotocous' mammals). The gestation period of rats, for example, is only about three weeks,

Table 1.2 *Gestation periods*

Species	Gestation period (days)	Comment
Mare	336 (329–345)	Roughly 11 months (longer with twins)
Cow	283 (270–291)	Just over nine months
Sow	113	Three months, three weeks and three days
Ewe	150 (143–154)	Roughly five months
Goat	148	Just under five months
Bitch	62 (60–63)	About two months
Queen	63	About two months
Rabbit (Doe)	32	About one month
Rat	22	Three weeks
Guinea pig	68	Just over two months (young are born in an advanced state of development)
Camel	395 (390–400)	13 months
Elephant	630	20–21 months
Rhinoceros	500 (480–520)	Approximately 16 months (black rhino a little less)

that of rabbits about a month, of domestic cats and bitches about two months and sows between three and four months. Each of these animals carries several young, contrasting with sheep and goats, whose gestation period is nearer five months. Cows and mares normally carry one offspring at a time (occasionally having twins), and their gestation period is 9–10 and 11 months, respectively. The elephant, rhinoceros and camel have longer periods still (see Table 1.2). The biological norm for our own gestation period is around nine months. Thus, the period of gestating foetuses is inversely related to the number that are normally produced, and also, it seems, is often directly related to the overall size of the animal.

Once a female is pregnant, there is no purpose in mating, so it does not usually occur. When the period of gestation is long, a female can direct a great deal of her energy into the growing foetus, and this is what happens in larger animals that carry fewer young. After giving birth, mares do not usually come into heat again for about three weeks or more. This gives them a bit of a breather, before expending more energy on the next pregnancy. Cows are not usually mated for about two months after calving, and sows not until at least a week has elapsed after the piglets have been weaned. Apart from the situation in horses, this is usually more a matter of the choice of the breeder than a natural biological response, but it still allows a little energy to be conserved or regained. If we exclude pigs, we can say that most of the polytocous mammals, with their relatively short gestation periods, adopt a different sort of tactic to give themselves a rest from pregnancy. This is to breed only seasonally. Monotocous sheep and goats also use this device.

Some domestic animals such as the cow and the sow appear to have abandoned seasonality and can breed all the year round, but seasonality, or at least intermittent breeding, seems to typify the majority of mammals. Horses, for example, are capable of breeding throughout the year, but the horse breeding fraternity would, I think, mostly concede that an element of seasonality comes into equine reproduction, over and above human choice. In the wild also, some species of mammal have discarded seasonal breeding. The kangaroo is one such example, but it applies to other Australian marsupials too. This looks to be a form of adaptation to the unpredictable conditions of extremely arid environments that prevail in parts of Australia. Indeed, it is difficult to be convinced that any of the marsupials display true seasonal breeding, although there may be a hint in some of them, and one is bound to wonder if this does not also apply to South American marsupials. There could be doubt about the Australian marsupial mouse *Antichinus*, for example, but in this species, the males die shortly after copulation. It can hardly be denied that this is quite an efficient means of ensuring it doesn't happen too often! It gives less competition for the next generation of males too ('That's him out of the way!').

Other Australian mammals, including the platypus and echidna, are distinctly seasonal. With the exception of monoestrous species, smaller laboratory mammals, virtually all British wild mammals and many small mammals in Africa, Asia and the Antipodes are seasonal breeders, with wild ruminants being conspicuously so.

Most rodents such as ground squirrels, flying squirrels, chipmunks, muskrats, beavers and voles are seasonal maters. Gophers appear to be intermittent, but not seasonal, whilst the springhaas or African jumping hare is sexually active all the year round. Wild rabbits and hares, including the Arctic hare and the snowshoe hare of America, are strikingly seasonal; the testicles of the brown hare of Britain progressively increase in size between November and March, working up to a crescendo, as it were, for the mating season which ends in July or August. After this, the hare's testicles start to regress for two or three months before starting up again. Peak activity in both sperm production and male sex hormone levels, therefore, occurs in March in these hares, when they form into groups and bounce and spring about all over the place. They appear to go slightly dotty at this time and their behaviour has been appropriately labelled 'March madness'. It is undoubtedly a manifestation of a high level of male sex hormone activity.

In Britain, a variety of well-known carnivores have distinct mating seasons. These include weasels, stoats, polecats, ferrets and badgers. On the American continent, raccoons mate seasonally, and in Africa, civets are seasonal too. Opinion is less firm when it comes to some of the African shrews such as the elephant shrew, but it is likely to be seasonal since the east-coast shrew and the yellow-backed elephant shrew are. Details of the kinkajou (honey bear) are not readily available, but I would guess that it is seasonal. Giant pandas (and perhaps other pandas too) would appear to be monoestrous and produce a singleton (sometimes two) only in the springtime. If they are polyoestrous, the system would have to be extremely inefficient. But aspects of reproduction in the panda are something of a mystery – otherwise it would be much easier to breed them in captivity.

The great majority of mammals seem to mate only seasonally, however, and hibernators such as bats, bears, marmots and dormice are no exception. Marine mammals such as seals – including the great elephant seal and fur seals – and presumably also including the walrus and sealion, each have their mating seasons.

The seasons of mating are not standard and may be governed in part by location. Two major factors are the cause: the so-called *ultimate* principle and a number of *proximate* principles. These principles work as follows. The ultimate principle is associated with the timing of the birth of young. However long the gestation period may be, it is clearly important that, if the young are to maximize their chances of survival, they should be born at a time of the year that is conducive to that survival. Take European lambs as an example. If they are to be born in the spring, when the weather is less inclement than in the winter, and if gestation takes about five months, then mating needs to take place before Christmas. The mating season of most sheep in Europe, therefore, begins about October or November. This indicates that the ultimate principle is probably the most important influence on mating seasons being restricted to a particular time of the year.

In addition, however, proximate principles, such as rainfall, food supply and temperature often come into the picture, so that mating seasons, even in the same species, may differ according to where the animal is located. This is commonly the case in the wild.

Such a difference is graphically illustrated by the mongoose. In this animal, mating occurs between February and July in Hawaii, but is shifted forward to a much later time in Fiji. There is a small mammal in Africa and the Middle East called a rock hyrax or dassie (it is the biblical cony and we shall hear more about him later because, reproductively, he is extremely interesting). This little chap has a well-defined mating season in South Africa, but as latitude decreases and we draw nearer to the equator, his fellow hyraxes are seen to mate progressively later in the year.

In other mammals, especially the larger ones, sperm production might rise to a peak at certain times of the year, even though it is continuous. This is what happens in the elephant and the hippopotamus. There is a very distinct mating season in camels, yet the males also produce some sperms throughout the year. Among smaller

FIGURE 1.4 A rock hyrax peering over a ledge.

mammals, the grey squirrel does the same. Domestic rabbits breed all the year round, but the quality of their semen has been seen to decline in August and September. In the majority of species, however, males have become just as seasonal in their reproductive activity as females, because there is no point in chasing a female who doesn't want to know. It is a hopeless cause.

In yet other species such as the dugong or sea cow (a close relative of the manatee) and some primates (relatives of monkeys), sperm production is not continuous all through the year, but neither is it seasonally based. It means that in these animals, mating is also periodic but not seasonal. This is rather difficult to explain, but perhaps social factors and even vegetation, a proximate principle, may be cues ('Look, here is some new green food, let's mate!'). It could also be that changes in environmental temperature play a part – we do not really know for certain. We do know that in these animals (known as intermittent breeders rather than seasonal breeders), males and females consort with each other at certain times without mating and that there is a distinct circannual rhythm of testicular activity in the males. Thus, there is a temporal protection of gametes as well as a spatial one.

SEASONAL ACTIVITY OF THE TESTICLES

To try and judge when mating seasons occur in wild animals is not always easy. If the size or weight of the testicles are taken as a measure, the results can be misleading and are often unreliable, unless we can be absolutely certain that the animal under examination is not immature. Juveniles are difficult to recognize in many of the smaller animals, and to add to the difficulty, impaired or totally inhibited sperm production with correspondingly small testicles in one individual, or even in a few, does not necessarily indicate a non-mating season in the population of the group as a whole. Patterns of social behaviour in a group may have a major influence on how active the testicles of any individual within that group may be.

On the other hand, among animals that live in groups, surplus males or those vanquished in a fight for dominance may leave and form bachelor groups or herds. We see this in antelopes (very noticeably in the springbok), and in the rock hyraxes of Africa and the Middle East. In these bachelor animals, sperm production is usually brought to a complete or almost complete halt, and levels of male sex hormone fall sharply. It is a sort of social castration, and is just as well as otherwise these males would be feuding the whole time instead of living amicably together as they do. Sometimes, one or more of these bachelors becomes sexually active again, but then he joins the females again. Others, probably mostly older animals, may leave the bachelor group only to wander alone and remain sexually inactive for the rest of their life.

Among male rock hyraxes, sexually inactive animals are sometimes to be found in groups in which most of the males are sexually active. Differences in sexual season have been recognized in different hyrax groups in the Rift Valley where many of them live, but variations within a group are more surprising. It is possible that individuals that show this peculiarity have their own personal testicular rhythm of sperm production that is unrelated to season, but if this is true, it is even more difficult to understand. Social factors might come into it, although we really don't know what the answer to this riddle is. But again, we have to be sure that these animals are not simply immature.

A number of animals that live in groups and have a hierarchical social structure mate for reasons other than reproduction. This probably includes some of the primates, but it is known that porcupines, for example, mate for purposes of family cohesion. Thus, mating times cannot be considered as clear-cut.

Obviously, the control of sperm production by a testicle is a complex business, but environmental factors play a major part in it, and day length, in particular, is crucial to defining most mating seasons. However, different animals respond differently even to changes in light (day length). In Europe, for example, rams and billy goats become sexually active when the days grow shorter at the back end of the year. Thus, they are called 'short-day breeders'. They are in complete contrast to 'long-day breeders', whose testicles become fully functional in the spring when the days are getting longer. A great deal of research has been carried out on this subject in hamsters and rams, because both are very sensitive to changes in light. Unlike rams, however, hamsters respond to a change from dark to light, so they are classified as long-day breeders.

The pineal gland, which is a latter-day third eye, just a remnant (although possibly functional in some fishes such as the lamprey and maybe even in some reptiles), is situated just between the two cerebral hemispheres of the brain. It plays a vital role in the response of testicles to changes in light at different times of the year. We refer to these changes as 'seasonal photoperiodicity'. We are not sure exactly how the pineal operates in this respect, but it responds to light through the medium of the eye, and when stimulated it releases a hormone called *melatonin*. This substance can exert an effect on the synthesis and release of testicle-stimulating hormones from the base of the brain (the pituitary gland) and thus indirectly turn the testicle on or off. It has been suggested that the testicle itself can respond to changes in the photoperiod, but this needs more evidential confirmation.

In most birds and also in our old friend the rock hyrax, the restoration of activity within the testicle (recrudescence) after a period of sexual quiescence is quite dramatic, since the organs show enormous growth. At the height of the mating season in hyraxes, believe it or not, the testicles come to fill almost the whole of the abdominal cavity, displacing other abdominal organs. Testicles in this species, you see, are situated in the abdomen rather than in a scrotum. Under the circumstances, it is fortunate that they are! At mating time, the testicles are very much the dominant organ within the abdomen. When the mating season is over, the testicles of these animals regress and become remarkably small (sometimes not much larger than a broad bean). A similar situation occurs in hedgehogs, so in this respect, these species resemble seasonally breeding birds. These are extreme examples of changes in testicular size in the mating season. Rams' testicles firm up in the breeding season, while a more marked and obvious increase in testicular size is seen in some of the antelopes when they are in rut, something also seen in hares in the spring. It is probable that a slight increase in the size of the testicles as they recrudesce from quiescence is typical of mammals as the mating season approaches, but it is more obvious in some species than others. After all, the mating season is the time when most sperms are needed, and the more the merrier as far as reproductive fitness is concerned. Large testicles, generally speaking, signal a higher level of sperm production.

Since ultimate principles play such an influential role in mammalian reproduction, it is clear that mammals invoke what is called a K-strategy of reproduction, in that they plump for survival quality in their young rather than concentrating on large numbers. Birds do the same. Thus, both mammals and birds bear fewer young or lay fewer eggs than, say, reptiles or amphibians, in particular. These other creatures adopt an r-strategy, which is to produce a large quantity of eggs and sperms as an insurance policy for reproductive success. Several fishes and molluscs, such as the squid, use the same strategy.

The mammalian approach depends on there being some stability in the environment at the time of birth, whereas r-strategy animals, such as some reptiles, frogs and other amphibians, seem to anticipate and usually experience environmental caprice. As a result, there is a lot of wastage of eggs, so they need the original product in plenty. Newborn turtles are vulnerable to predators (all part of the environment), but a lot of eggs are helpful for the survival of a few. The system can be very successful, as witnessed by the alarming spread of cane toads in Australia. The toad, *Xenopus*, is beloved by many biologists because it produces thousands of eggs for them to study. No mammal can match that, and only the elephant shrews of Africa, which ovulate a hundred or more eggs at a time, even try.

Clearly, it is an uphill struggle for a male mammal to copulate and disperse his genes, when other males and all females seem to expend considerable energy in trying to prevent him from doing so too often. This is the system, however. The following discussion looks at how the males of different species try to overcome the system through aspiring to reproductive fitness and how some of them eventually prevail.

FURTHER READING

Asdell, S.A. (1965) *Patterns of Mammalian Reproduction*. Constable, London.

Austin, C.R. (1965) *Fertilization*. Prentice Hall International, Inc., London.

Birkhead, T.R. (2000) *Promiscuity: Evolutionary History of Sperm Competition and Sexual Conflict*. Faber & Faber, London.

Birkhead, T.R. & Brennan, P. (2009) Elaborate vaginas and long phalli: post-copulatory sexual selection in birds. *Biologist*. **56**, 33–38.

Borradaile, L.A. & Potts, F.A. (1961) *The Invertebrata*. 4th edition. Cambridge University Press, Cambridge and New York.

Brennan, P.L.R., Clark, C.J. & Prun, R.O. (2009) Explosive eversion and functional morphology of the duck penis supports sexual conflict in waterfowl genitalia. *Proceedings of the Royal Society, London. Series B*. **10**, 1–6.

Cohen, J. (1977) *Reproduction*. Butterworths, London, Boston, Sydney, Wellington, Durban and Toronto.

Cohen, J. (1999) The evolution of the sexual arena. In: *Male Fertility and Infertility.* Eds: T.D. Glover & C.L.R. Barratt. Cambridge University Press, Cambridge.

Dawkins, R. (1983) *The Selfish Gene.* Granada Publishing Ltd, London, Toronto, Sydney and New York.

Dawkins, R. (1988) *The Blind Watchmaker.* Penguin Books, London, New York, Victoria, Ontario and Aukland.

Dawkins, R. (1996) *Climbing Mount Improbable.* Viking, London, New York, Victoria, Ontario and Aukland.

Dawkins, R. (2009) *The Greatest Show on Earth.* Random House Ltd, London and New York.

Dukes' Physiology of Domestic Animals (2004) 12th edition. Ed: W.O. Reece. Comstock, Ithaca, NY.

Hearn, J.P. (1982) The common marmoset (*Callithrix jacchus*). In: *Reproduction in New World Primates.* Ed.: J.P. Hearn. MTP Press Ltd, Lancaster, Boston and The Hague.

Jewell, P. (1977) The evolution of mating systems in mammals. In: *Reproduction and Evolution.* Eds: J.H. Calaby & C.H. Tyndale-Biscoe (Proceedings of the fourth Symposium on Comparative Biology of Reproduction held in Canberra in 1976). Australian Academy of Science, Canberra.

Maynard Smith, J. (1978) *The Evolution of Sex.* Cambridge University Press, Cambridge, New York and Melbourne.

Millar, R.P. & Glover, T.D. (1970) Seasonal changes in the reproductive tract of the rock hyrax (*Procavia capensis*). *Journal of Reproduction and Fertility.* **23**, 497–499.

Millar, R.P. & Glover, T.D. (1973) Regulation of seasonal sexual activity in an ascrotal mammal, the rock hyrax: *Procavia capensis. Journal of Reproduction and Fertility: Supplement.* **19**, 203–220.

Neaves, W.B. (1980) Asynchronous testicular cycles among equatorial colonies of rock hyrax (*Procavia habessinica*). In: *Testicular Development, Structure and Function.* Eds: A. Steinberger & E. Steinberger. Raven Press, New York.

Roth, T.L. (2006) Review of reproductive physiology of rhinoceros species in captivity. *International Zoo Yearbook.* **40**, 140–143 (this paper is essential for those with a particular interest in breeding rhinoceroses).

Skidmore, J.A. (2005) Reproduction in the dromedary camel: an update. *Animal Reproduction.* **2** (3), 111–117 (this is an excellent paper for anyone interested in camel reproduction).

The preamble

It is easy to imagine that to achieve his goal of copulation, a male simply has to follow three easy steps. First, make your presence felt; second shoo off all competitors; and third, persuade your female or females to submit to your blandishments. This is basically what happens in birds, but mammals, as we shall see, are slightly different.

Male birds attract the attention of females by beautiful, and sometimes not so beautiful, love songs. When, after this, a female comes into the male's space, he shows off with magnificent crests and tails. Just picture some of the birds of paradise, the humming birds, lyre birds and other birds of Asia, the Antipodes and the Americas, displaying the exquisite hues of their plumage. Perhaps best known is that giant of show-offs, the peacock, with his psyche-delic tail feathers glittering in the sunlight. Oh yes, he can hardly help but catch the female eye.

Surprisingly, perhaps, in view of such displays, we see in a number of bird species several males mating at different times with a single female (*polyandry*). A famously polyandrous male bird is the bower bird of Australia, who invites his females to visit his den (or bower) by adorning it with bright and often gaudy colours. He does so by stealing anything around that shines, such as pieces of rock, silver paper or even buttons. Nevertheless, a female bird can be quite particular about whom she accepts. The male pheasant, who typic-ally mates with a number of different females, resplendent in his rich and contrasting coloured feathers, stalks his less prepossessing female, but it is she who does the choosing. This is probably true also of birds such as robins and some of the penguins, who are supposed to have a single mate if they play it by the book (but it doesn't always work out that way!). One cannot help but wonder

FIGURE 2.1 The peacock is a great show-off! Photograph © iStockphoto. com/Naomi Woods.

though, just how choosy those female birds are that copulate with several males!

Outside the range of such mammals as the chimpanzee (which, we will see later, is promiscuous), multiple copulations by different males with a single female are not usual in mammals, and when it does happen, it is rather by accident than design and rarely if ever involves more than two males. Bitches may copulate with more than one dog, and it has been known for many years that they may even have pups in the same litter that have been sired by different dogs. This, however, is definitely not the mammalian rule, although the egg-laying echidna, like a bird, goes in for polyandry.

But different males of a single species are competitive and this is to be seen typically throughout the animal kingdom. Elegant work on dungflies opened up the concept of 'sperm competition'. The relatively rare double matings in mammals have indicated that it can be the second copulation by a separate male that is successful, rather than the first. It looks, in these cases, as if sperms do compete for success, and the whole idea has given us all pause for thought and cause to revisit earlier conventional concepts. I am not sure that the term is the best one, but this is purely personal and I cannot think of a better one. Certainly, the advent of oestrus, unique to female mammals, is an intrusive factor in applying the idea to mammals generally, since ultimately there can be little or no female selection. There are also plenty of systems (Chapter 5) for keeping the semen of second or third males at bay, although these could be construed in themselves as being competitive or selective principles.

Like birds, male mammals have different sexual strategies. There are those that consort with one female, or two at most, either for life or throughout a mating season. These are *monogamous* males. Monogamy is comparatively rare among mammals, but otters are monogamous, as are prairie voles in America and some of the smaller African antelopes, which are known to form pair bonds. Among primates, the gibbon and the lemur both form life-long relationships with females. By contrast, *polygamous* males mate with several females within a group. Some antelopes, for example, form harems. Occasionally, an apparently wayward female will wander off and find herself in another's harem. This also happens in strictly territorial (but not polygamous) seals. Straying into another's territory could be more by accident than a design of infidelity (at least, this would be her excuse!), but there is evidence to suggest that these perambulations may not always be unintentional. If this is true, then these particular females are definitely behaving like birds! *Polygynous* animals (including seals) are those in which the male copulates with more than one female, but is not especially competitive once he has claimed them. He may be protective, but has competed for dominance before he has acquired his females. Some polygynous males may have a host of females, like the ram and the stag or buck deer; others, such as the great apes have fewer, but more than one, of them around him at a time. *Promiscuous* males mate with any female who happens to be passing by. By this I mean *any* female, whatever her age, shape or size, provided she is willing. The chimpanzee is a prime example.

These species-specific stratagems are not only related to the continuation of that species, but also to the fitness of its population. They may have evolved, of course, in relation to social structures. In evolutionary terms, Darwin contrasted *natural selection*, which is dependent upon the success (i.e. the survival) of both sexes, and *sexual selection*, which concerns the success of individuals of the same sex over others. The first type of selection, therefore, relates to the success of a population, whilst the other, as Darwin pointed out, is a question of a sexual struggle. Sexual rivals need to be driven away before a charm offensive on females, just to make sure they are receptive, can begin. This applies to all sexually reproducing animals, but they all do it rather differently. It is evident, for example, that birds and mammals have taken a rather different path in this respect over the course of their evolution. This most likely stems from the fact that they have evolved differently and have different social systems, but it means that they cannot really be compared, because only in mammals is oestrus among the females a characteristic feature.

If a male mammal is to prove his reproductive fitness, he needs to be accorded appropriate weaponry to establish his dominance over other males. The main weapons used, it seems, are those of size and strength. In most mammalian species, males are larger than females, because it is they, not the females, who have to see off sexual competitors. It is particularly well demonstrated by that mighty polygynous giant, the gorilla. The male orang utan is also twice the size of the female, whilst the polygynous male seal is eight times the weight of a female. This discrepancy in size is known as *sex dimorphism*.

By comparison, since male chimps are less competitive, if at all, there is little to choose between the size of males and females.

Curiously, the female can sometimes be bigger than the male. Female rabbits and hares are larger than the males, as are hamsters and some of the bats, including the common bats, the long-eared bats and the pipistrelles. It occasionally happens among the antelopes too. But the most dramatic example of female superiority in size is the spotted hyena. It is not too clear how this sort of reversal happened, and it might be of variable origin. It certainly cannot be a matter of sexual selection as it is in the male. As far as the spotted hyenas are concerned, it looks to have a social basis. Unlike their cousins, the striped hyena, spotted hyenas have very elaborate greeting rituals and it is the dominant female who does the sniffing around when meeting another pack. This could be a sensible development, because if males were to do it, they might kill each other! Interestingly, though, all female hyenas have a protuberant vulva, which makes them look rather like males. This led to a belief among some African people that the hyena could change its sex at will and that this is why it is always laughing! This is not quite as comedic as it at first sounds, because female spotted hyenas do at first develop male external genitalia.

However, sex dimorphism in favour of males is generally the rule among mammals. But size and strength may not always be enough to ensure your superiority; other weapons have made their appearance in some species.

SEXUAL WEAPONS

Take horns and antlers as examples of these additional weapons. When the rut is on, a deer stag or buck sports his notoriously ornate antlers, long after they have shed their velvet. What a splendid sight the 'monarch of the glen' (red deer stag) is! In addition to his magnificent antlers, he stands there on the mountain top with stout neck, swollen throat and enviable pectorals. But to his sexual opponents he is intimidating, even deadly. The bigger the antlers the better

FIGURE 2.2 'Challenge me if you dare!' Monarch of the Glen. A red deer stag in rut. Photograph © iStockphoto.com/Damian Kuzdak.

when locking horns with an intruder, and the more ample the branches, the more likely you are to hold him in a lock and the better the chance of lancing him with a piercing and fearful stab.

Clashes between stags are so ferocious, it is surprising that more antlers are not broken. But they are tough, and like all animal horn, are possessed of great tensile strength. Veterinary surgeons who have de-horned adult bulls will testify to this, as will many a matador who has had the full force of a horn in his groin. Just think, he cannot walk away or turn his back either, lest, with equal devastation, he gets it up another part of his anatomy! When male gnus (wildebeests) in particular, but even horned rams, mountain goats, hartebeests and all the male antelopes, crash into each other in sexual conflict, we can hardly help saying 'Ouch!'. These animals certainly mean business, but it is not usually a fight to the death. It is simply a battle to show who is the boss. Sometimes, it gets out of hand and death of the vanquished combatant ensues, but this is accidental. Usually, the loser soon gets the message and either sprints or limps away, according to what point has been reached in the encounter.

Competition is now ended and, especially in polygynous males, it is necessary that it should be. A ram, for example, may copulate with 30–40 ewes in a day, so he cannot waste time fielding competition. Of course, cheeky young aspirants may try to sneak into the flock when no one is looking, as it were, but a watchful eye of a dominant male will usually spot him and he will pay the penalty for the error of his daring (even if he might have a quick fling before being caught). Lions and other male cats behave similarly.

If a male has no horns, there are other instruments of war available. Hornless rams such as Suffolks have what seem to be armour-plated skulls that slam into an opponent's head. Rearing stallions flail their competitors with foreleg hoofs. Hares and kangaroos, in their face-to-face confrontations, can deliver an extremely bruising kick from the hindlegs through leaping in the air and aiming at an opponent's chest or abdomen. Bears snarl and grapple. There is the all-in wrestling of the lions of Africa or the tigers of India, with teeth bared and claws unsheathed and extended; then there is the menacing gape of competing male hippopotamuses with their massive tusks. The great male elephant seal has enormous weight and size and he hurls his gigantic chest forward with reverberating thuds up against his opponent. Similar activity may be seen in the head-lunging of male whales, but elephant seals also use their highly developed canine teeth as weapons, and so frequently the blood of a sexual competitor flows freely.

All this behaviour is stirred up by an increase in the level of male sex hormones in the bloodstream, and perhaps increased receptivity to the hormones of sexual organs, including the brain. Collectively, male sex hormones, and there are several of them, are called *androgens*, but the leading one and most potent of all is *testosterone*. And powerful stuff this hormone is. I once separated a red deer stag from the hinds in my herd by a large and substantial padlock on a

FIGURE 2.3 Male elephant seals confronting each other. Photo © iStock-photo.com/Dave Adalian.

strong metal gate. When he came into rut, he smashed the padlock open and mangled the iron gate as if it were made of papier mâché. He reached his hinds! In elephants, too, it is testosterone that is responsible for *musth*. When this comes on, Indian elephants have been known to break their chain tether, activate secretion from glands at the side of the head (temporal glands), which are controlled by testosterone, and generally go quite berserk. This has given rise to the expression 'running amok', (the word 'amok' being a Malayan Hindu word). Levels of circulating testosterone during these phases can be astronomical. The phenomenon is entirely separate from seasonal breeding and seems to be nothing to do with a desire to mate.

Quite a number of mammals also have scent glands which have potential in defence or even offence. The skunk is perhaps the nastiest of all in using this pungent form of defence or aggression. Although it probably doesn't use it in sexual confrontations, it is very useful as a territorial marker. Dogs have rather useless anal sacs, which indicate that their forebears might also have used smell, if not as a weapon, as a territorial marker. Such markers are also testosterone-sensitive.

TERRITORY

The marking out of territory is important, not only in polygynous ruminants (herbivores with more than one compartment to their stomach, such as the huge Eland bulls of Africa or smaller antelopes), but also in pack animals such as wolves, wild dogs and related species. When a domestic dog cocks his leg against a lamp post, he not only discharges urine, but prostatic fluid as well. Dogs have a large prostate gland and produce as much prostatic fluid in an hour as a man does in a day. He has to get rid of it somehow and this is how he does it. Larger animals do the same. When a rhinoceros or a hippo-potamus sprays, it is quite something!

Territorial marking may announce a male's presence, but most likely, it is primarily intended to warn off competitors by saying 'This is my patch. Keep off.' When cats rub their faces against an owner's leg, as an apparent gesture of love, it is probable that it is more a matter of leaving their scent. Deer mark out their territory by using glands that are situated below the inside corner of the eye (the medial canthus) and can thus be seen rubbing their faces against bushes. This is their equivalent of cocking their leg to leave their card. It has been claimed that with antelopes, territory comes first and sex afterwards. I cannot agree with this view. The more forceful argu-ment in my submission is that reproduction is the prime imperative and that territorial claim is simply a part of it. Without reproduction, there would be no territory to claim! But territoriality is an interest-ing phenomenon and, incidentally, is a prominent need in man.

Rock hyraxes have a peculiar dorsal gland on their back, which expands and fills with fluid during the mating season. It does not appear to secrete the fluid externally, but it may be a signal to other males in the group that they may come close for collective protec-tion, but not too close. This gland makes sexually active males more

FIGURE 2.4 Typical grouping of rock hyraxes to maximize breadth of vision. Note the dorsal gland on the back of each animal.

visible to aerial predators, but rock hyraxes collectively fan out on the top of rocks, which increases their overall angle of vision. They are also very agile and a message of danger is likely to be passed quickly through the group, so that in a flash they scatter into the safety of various holes and crevices. This gland, like the scent glands and the prostate in other species, is under the control of testosterone (see Figure 2.4).

LOVE CALLS

Sexually active males tend to be more noisy than usual, but in mammals, unlike birds, the calls are not so much to attract females to come to the male, but rather indicate that he is in the vicinity and intends to come to them. It calls their attention to his presence and, above all, warns off other males who might have designs on females in the neighbourhood. There can be a bit of beckoning by the female, such as the low sound of a female elephant when she is in heat. We presume it means 'come closer'; similarly, the female orangutan might screech and bang her breast when fully in oestrus. However, when we hear the haunting call of the stag on the hillside, we know that he is ready for action and that sex is in the air. The thrilling roar of a lion in the African night – and what a wonderful sound that is! – lets us all know that he is on the prowl and that any would-be competitor had better steer clear. The eerie high-pitched wail of a wolf in the snow-capped woodlands (often backed up by the whole pack joining in the song), the gritty gurgle of a howler monkey in the tropical forest, the stamping foot of a buck rabbit or hare, the hoarse low nicker of a sexually excited stallion and the bellowing of a bull all send similar messages and announce that high levels of testosterone have arrived again. Now a female or a band of females looms into view, so what is the next move?

FOREPLAY

Having claimed victory over his opponents, and in some cases having made clear the boundaries of his territory, a male mammal may now get down to softening up his female or females. As he approaches, he needs to ensure that she is ready for him. The process of coaxing or wooing differs with different mating systems. As with rams, it is brief in polygynous antelopes, but related animals may indulge in a spot of foreplay and may be seen stomping or pronking (prancing on all fours), pretending to draw close, then dashing off as if to tease her (although she can do this too, as though to egg him on). When, eventually, he considers her to be fully in heat, he may nibble her neck, rest his chin on her back and generally caress. An experienced stallion will play with a mare for quite some time before he mounts her, just to ensure she is fully in heat. A mistake could lead to a good kicking. Sometimes a vasectomized horse of lesser merit is used as a 'tease' to make sure this doesn't happen. When a mare is fully receptive, the clitoris is erected periodically and she is said to be 'winking' (Figure 2.5). In the wild, the way stallions avoid kicking mares and yet still keep chasing them – through deftly turning their head at crucial moments – can really be quite impressive. Przewalski stallions

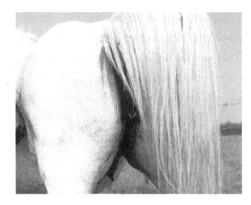

FIGURE 2.5 A mare in oestrus 'winking'. Courtesy of W.R. Allen.

(Mongolian wild horses) have been caught on camera displaying this behaviour. Only perhaps when she tires can a stallion determine with certainty that the mare is fully in heat and be confident enough to draw close. She, too, might be impressed by his persistence, but she cannot help herself once she is fully in heat.

FIGURE 2.6 A male giraffe in caressing mode.

This stage of mating can sometimes be quite touching to observe. The quiet intertwining of necks by giraffes is very graceful, and that giant of the ocean, the whale, moves close to his female with considerable gentleness, as does the huge bull elephant. Even the rhinoceros, who treats his female roughly, ends on a softer note immediately prior to mounting. At this time, the female needs reassurance, so the male has to be courteous. In mammals there is none of the ducking, bobbing, bowing and paddling rituals that we sometimes see in birds, because the male is not trying to attract his female, he has already got her.

PHEROMONES

The subject of male smell is dealt with more fully in the section on human reproduction, but it may also play a part in animals. The smell of billy goats is unmistakable, and rams have specialized sweat glands in their nether parts. The scrotum, too, can be fairly smelly. There may be nothing like sticking your backside under the nose of your female for making her release her eggs! A few decades ago, a wonderful scientist called Dr Hilda Bruce showed that in mice the very smell of a male was enough to terminate a pregnancy. Presumably, the spontaneously aborting females wanted to mate and only by this means could they ensure that they came into heat. This sexual smell is caused by substances known as *pheromones*. The response of the mice was named 'the Bruce effect'.

Male smells are not always welcomed. In Soay sheep (a rare British breed), the smell of the ram is so obnoxious, the ewe will not let him near her again once he has briefly mounted her and ejaculated. It has been reported that a second male might try, but it would be surprising if he succeeded, if the stench of the first one still remains! It certainly seems like a good scheme for making sure that polygyny works!

The take-home message here is that the mating game is a complicated business and that although copulation may be the ultimate aim, it is only a part of the whole process. Second, it has to be appreciated that, unlike birds where polyandry is relatively common-place, it is not so in mammals. Intersexual selection of males is, therefore, vital to the reproductive success of many birds, but intra-sexual selection is the more important in mammals. Male birds show off to their females, but typically, male mammals show off to other males ('I am the finest here, so make no mistake, unless you want a big fight on your hands.').

There is no firm evidence of cryptic female choice in mammals as there can be in birds. In fact, broadly, female mammals seem peculiarly unimpressed by male physical features, and if they are in heat they are impelled to copulate with any willing male, however puny. A female mammal usually copulates with a dominant (alpha) male, because he is the most likely one to be around to do the job when she comes into heat.

There has been a suggestion that the bright blue scrotum of the colobus monkey is a sexual display feature to attract females, but it is conjectural and probably not true, although when at its brightest, it might signal a willingness to mate. Nor do female mammals appar-ently show much enthusiasm for the act of coitus itself.

They show no fluttering ecstasy as in a bird. During its course, female mammals usually look as if they find the experience rather bleak, or at least slightly tedious. Mares and their relatives, for example, stand with laterally flattened ears as if to say 'Well, I suppose it has to be', a sort of equine version of 'Lie back and think of England' (see Figure 2.7). We cannot be absolutely sure that female mammals do not enjoy coitus, and interesting work is being done on pituitary activity (see Chapter 3) and the release of dopamine (the feel-good hormone), but this hormone relates more to bonding than to the actual joys of coitus.

As far as can be seen, large quadrupedal animals do not experi-ence female orgasm, which also applies to dogs and smaller animals. But to the male, it is not important whether his female enjoys coitus or not, that is not his aim. He has done his stuff and posted the mail. Nothing more need be said, because with a few exceptions, male

FIGURE 2.7 Coition in donkeys. Does she look to you to be enjoying herself? (See text). Courtesy of Lee Morris.

mammals play no part in the rearing of young. Famous exceptions are marmosets, notably the lion marmoset and tamarin monkey. The males of these primates are not only supportive of the female in pregnancy and at the time of the birth of her young, but they also care for the little ones after they are born, handing them over to the female only for feeding. It is conceivable that they might even lactate themselves! Otherwise, and typically, it is fair to say that male mammals are not geared to rearing young.

Sometimes, it seems that sexual adornments outgrow their usefulness. The antlers of a deer, for example, can grow to be so heavy and fancy that they must be, like the peacock's tail, an encumbrance. This means that they become a detriment to the individual. This is a case of sexual selection taking precedence over natural selection. But what does it matter, if you have already made sure of a future generation of potentially fertile (reproductively fit) males and sexually active females? Male mammals rarely rule for life, and with younger and more agile animals around the place, it is the urge to copulate that has top priority. In lions the drive can be so strong as to lead to infanticide. When most female mammals are suckling their young, they cannot come into heat, because lactation inhibits oestrus. If the young are removed, however, milk production will end and an oestrous cycle will return. If a male lion kills the cubs, therefore, and it has been recorded, he will be able to copulate and deliver more of his genes. This underscores the reality: the male sexual urge can be alarmingly forceful. Such infanticide is a danger in other big cats and also among bears and other species.

To achieve satisfactory copulation, which we have seen is the most important ambition for a male, he must be endowed with appropriate equipment to do the job. A discussion of this equipment is next on our agenda.

FURTHER READING

Ardrey, R. (1961) *African Genesis*. Collins, London.

Ardrey, R. (1969) *The Territorial Imperative*. Collins Fontana Library, London.

Bruce, H.M. (1970) Pheromones. *British Medical Bulletin*. **26**, 10–13.

Glickman, S.E., Short, R.V. & Renfree, M.B. (2005) Sexual differentiation in three unconventional mammals: spotted hyaenas, elephants and wallabies. *Hormones and Behavior*. **48**, 403–417.

Lincoln, G.A. & MacKinnon, C.B. (1976) A study of delayed puberty in the male hare *Lepus europaeus*. *Journal of Reproduction and Fertility*. **46**, 123–128.

Lincoln, G.A., Guiness, F. & Short, R.V. (1972) The way in which testosterone controls the social and sexual behavior of the red deer stag (*Cervus elaphus*). *Hormones and Behavior*. **3**, 375–396.

Packer, C. & Pusey, A.E. (1983) Adaptation of female lions to infanticide by incoming lions. *American Naturalist*. **121**, 716–728.

Parker, G.A. (1970) Sperm competition and its evolutionary consequences in the insects. *Biological Reviews*. **45**, 525–567.

Sadleir, R.M.F.S. (1969) *The Ecology and Reproduction of Wild and Domestic Mammals*. Methuen, London.

3 The equipment and the product

That mammalian embryos start out essentially as females was rather startling news. But at first, the embryonic gonad (potential testis or ovary) is undifferentiated and if nothing were done, most mammals, including ourselves, would each develop into females (female elephants and, as we have seen, hyenas, by contrast, at first display male features, but they are exceptional). Usually, something is done, so that some originally undifferentiated embryos develop into males. This is because they become endowed with a testis. In other words, in these embryos, the gonad develops into a testis rather than an ovary. Thus it is that a testis normally casts the seal of masculinity on an individual, way back in its embryonic state, and if the gonads are experimentally prevented from developing, that is, an embryo is rendered agonadal (this can be done experimentally in rats by removal of the pituitary, a procedure known as hypophysectomy), it automatically proceeds to be female.

Precisely how the message is passed to make an embryo become a male is complex, but in mammals, it depends on the presence of a Y chromosome. Individuals that are destined to become females inherit two X chromosomes (XX), but males replace one X with a Y chromosome (XY). This depends on whether an X or a Y (female or male) sperm (with the haploid condition) penetrates and fertilizes the egg. X and Y chromosomes are called *sex chromosomes* to distinguish them from *somatic chromosomes*, which are associated with all other cells of the body. So as not to have too many X chromosomes (that is, one more than males), females eventually lose one so that each sex ends up with the same complement of X chromosomes. There is a bit of toing and froing between chromosomes too, whereby genes can occasionally stray from their original chromosome onto another one (translocation). This can even happen from a sex chromosome onto a somatic (body) chromosome, which itself has nothing to do with sex. As a result, so-called sex-linked genes occasionally arise on body chromosomes. Things are not as straight forward as they might at first seem to be.

Whole texts could be and have been written on the intricacies of sex determination, but suffice it to say here that the Y chromosome is the key to maleness and normally it ensures that a testis develops. Special cells within a developing testis, the Sertoli cells (which will be mentioned later in relation to the adult animal), play a major role in this development, even though they themselves are somatic cells in origin. Testosterone is soon produced by new and different cells within the developing testis and thus the male fate is sealed. If all goes according to plan, there is no going back and the basic femaleness of the embryo is suppressed.

Testosterone is responsible for the development of male genital organs, both internal and external, as well as influencing the brain and giving rise to other secondary sexual characteristics. The outcome of these events is that the framework for male reproductive activity is in place at birth and later springs into full action with the onset of puberty. It is then that the production of sperms gets into full swing.

The male reproductive product is semen (or 'sperm', sometimes referred to as seminal fluid) and it consists of actively motile sperms (spermatozoa) suspended and swimming in a fluid (seminal plasma).

THE MANUFACTURE OF SPERMS
Sperms are produced in the testes, which are two balls of tightly bound undulating tubules (*seminiferous tubules*) contained under considerable pressure by a tough inelastic tunic. This tunic (*tunica albuginea*) sends leaves (laminae) or strips of connective tissue into

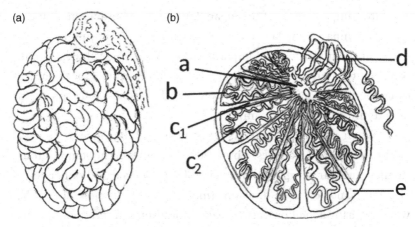

FIGURE 3.1 The testis. (a) The testis stripped of its outer tunic. This has to be imaginary, because the tubules being held by the outer tunic are under pressure, which means that if it is slit, the seminiferous tubules herniate. Thus, a ball of tubules without the tunic can never be observed. Some testes will have fewer tubules than this, and some testes (e.g. human) are lobular, so this line drawing can be nothing more than an indicator of what it might look like, if we could see through the tunic. Probably the arrangement of the tubules may be more regular in some species too. (b) Section through a testis to show its basic structure. Labels: a = rete testis; b = seminiferous tubule (tubular compartment); c_1 = connective tissue laminae extending from the outer tunic, which join to form the mediastinum (see text) and are part of the interstitial tissue, including the Leydig cells (intertubular compartment) – see also c_2 (another part of the interstitium); d = efferent ductules leading into the epididymis; e = tunica albuginea.

the centre of the organ, where they converge to form a central plank of tissue called the *mediastinum* of the testis (see Figure 3.1b). The tubules are held together, rather like egg binding the meat in a rissole, by interstitial tissue. This also consists of the connective tissue, but it has blood vessels, nerves and connective tissue cells in it as well. Most important of all, this interstitial part of the testis contains the *Leydig cells*, which elaborate testosterone and other androgens. Thus, the testes serve two functions: one is to produce sperms; the other is to yield male sex hormone. But each function is carried out in separate compartments of the organ. Sperms are manufactured in the tubular compartment of the testis, whilst male sex hormone is produced in the intertubular compartment (between the tubules). So, in addition to yielding sperms, the testes can veritably be regarded as endocrine organs (organs that produce hormones and pass them directly into the bloodstream).

When a male individual is sexually active, the process of sperm production (*spermatogenesis*) proceeds apace within the seminiferous tubules (tubular compartment). It involves a special and rather complicated kind of cell division (*meiosis* or *reduction division*), whereby, as touched on earlier, the chromosome number of sperms (and also of eggs in the female, since ovarian cells do the same thing) is halved, so that a sperm is said to have a haploid number of chromosomes. The unison of two haploid cells (sperm and egg) at fertilization, therefore, restores ploidy and the offspring or zygote is, in this way, possessed of a normal diploid set of chromosomes. In some invertebrates such as moths (*Lepidoptera*), meiosis does not occur until after fertilization, but the end product, namely a diploid zygote, is the same as in mammals.

At the periphery or outer rim of the thick-walled but essentially hollow seminiferous tubules lie the peculiar Sertoli cells, and between them is a row of stem cells or basal cells, which ultimately, through a series of cell divisions, give rise to sperms. These are the *spermatogonia*. There are two types: Type A and Type B. Several Type As are produced by mitosis (producing two replicas or clones of the original one) before any of them become converted into Type B, and each time one of them divides mitotically it produces a resting Type A and an active Type A. The resting types may burst into action at a later time, although there is a lot of wastage due to *apoptosis* or pre-planned cell death. The dead ones are eventually engulfed by Sertoli cells. However, the remaining ones continue to divide by mitosis and produce more daughter cells, known as *primary spermatocytes*. As a result, the process of apoptosis is not detrimental because there are always plenty of live spermatogonia in reserve. Once a primary spermatocyte is formed, it too divides to give rise to two *secondary spermatocytes*, and there may be further apoptosis at this point in the process.

Secondary spermatocytes are the first cells to divide meiotically to yield cells called *spermatids*. They do so rapidly, so they are only transitory cells and sometimes difficult to see on conventionally stained microscopic sections. But spermatids thus have a haploid number of chromosomes and without further division give rise to the sperms (illustrated in Figure 3.2). Soon the nuclei of the spermatids elongate and the definitive shape of the sperm head appears. Then the flagellum or sperm tail grows out due to progressive elongation of one of the cell's organelles (tiny structures in the cytoplasm) and thus the beginnings of a sperm are to be seen (see Figure 3.3a).

However, several events occur during a prolonged early phase of the division of primary spermatocytes, before meiosis begins. First, maternal and paternal chromosomes that have like gene loci on the same DNA sequences (homologous chromosomes) move close together and pair up ('homologous pairs'), being joined together by a central body called a *centrosome*. They are then seen to split, first giving rise to a tetraploid condition. But before these chromosomal units separate in further mitosis, their chromosomal arms (chromatids) cross over at points along their length (chiasmata), so that genetic material from one homologous chromosome is exchanged with the other. This is important for ensuring that after the chromosomal split, each daughter cell (secondary spermatocyte) gets the same sequence of gene loci. This is also depicted in Figure 3.2.

Mitotic division, which occurs in most somatic cells as well as in spermatogonia, is much simpler. The homologous chromosomes just separate to give identical daughter cells. Taking man as an example, there are 46 somatic chromosomes in total, plus two sex chromosomes in the female, XX, or in the male, XY. The result of meiosis, therefore, is that spermatids become endowed with 23X or 23Y and accordingly and collectively produce both X and Y sperms (look again at Figure 3.2).

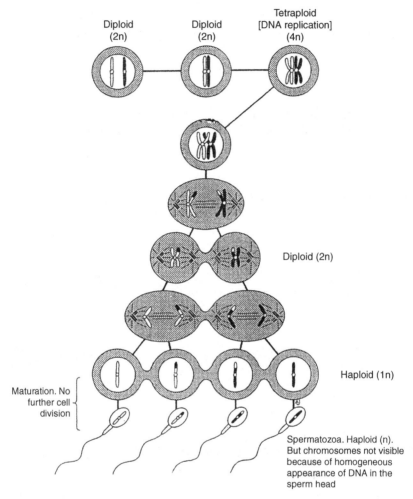

Diploid
(2n)

Diploid
(2n)

Tetraploid
[DNA replication]
(4n)

Diploid (2n)

Haploid (1n)

Maturation. No
further cell
division

Spermatozoa. Haploid (n).
But chromosomes not visible
because of homogeneous
appearance of DNA in the
sperm head

FIGURE 3.2 Cells associated with spermatogenesis, showing how the haploid condition arises in spermatids and sperms. Drawn by Ann Johnson.

The Sertoli cells have long finger-like projections and special junctional complexes between each other. So how is room to be found for the increasing number of cells produced by dividing spermatogonia, if they are locked in by these cytoplasmic projections of the Sertoli cells? Well, it must be that the Sertoli cell junctions give way and allow the increasing number of spermatocytes to fall through the gap.

The outcome is that each successive cell appears closer to the centre of the tubule – that is, the hollow part or lumen. So sperms (the end product) are mostly to be seen towards the inner part of the tubule, hanging on to Sertoli cells like bunches of grapes, with their tails pointing into the lumen. Thus, the seminiferous tubules appear to be lined by a thick wall of cells known as the *germinal epithelium* (Figure 3.3b). It consists of cells at different stages of spermatogenesis, and the inter Sertoli cell tight junctions divide it into what is referred to as a basal compartment (before the spermatocytes are released through the gaps) and an adluminal compartment, where meiosis occurs closer to the centre of the tubule (Figure 3.3a).

Spermatogenesis is a highly organized process, and in a microscopic section of a testis, any spermatogenic cell is always to be seen in association with one of its own kind. So, a particular picture is always presented according to which stage of the process has been reached. When the same cell associations are again seen in a tubule, it is said that a *spermatogenic cycle* has been completed. The time it takes for a cycle to be completed is constant for a species, but varies between species. When finished, the process begins again, but meanwhile, slightly further along the tubules, it will be at a different stage. For instance, it may be only half-way through the process. Typically, there are 14 different stages of cell association in a cycle (refer to Figure 3.3c).

Spermatogenic cycles occur in waves along the seminiferous tubule (*spermatogenic waves*). If you look at a microscopic section of tubules, therefore, you are likely to see different stages of spermatogenesis, according to which level of the tubule has been cut through. Each level of the tubule appears in a section to be a separate tubule, because the seminiferous tubule is undulating and several levels of the one tubule cannot be avoided in preparing a section for the microscope. But actual separate tubules may also be there, so interpretation of what is going on in a testis at any one time is not too easy, because some sections of tubule will appear to have the same cell associations, whilst others may be different (Figure 3.3b). To add

(a)

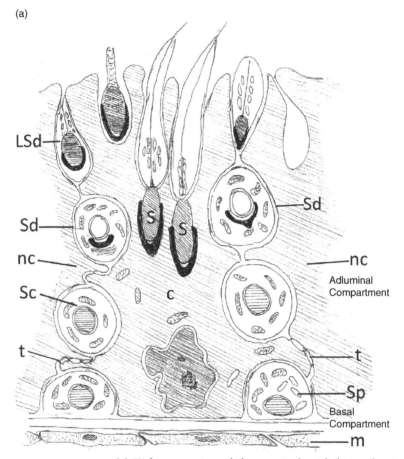

FIGURE 3.3 (a) High-power view of the germinal epithelium, showing how the Sertoli cell tight junctions divide it into basal and adluminal compartments (see text). Labels: m = myoid cells; Sp = spermatogonium; t = Sertoli cell tight junctions; Sc = spermatocyte; Sd = early spermatid; LSd = late spermatid; S = spermatozoon (sperm); c = Sertoli cell with nucleus and organelles such as mitochondria; nc = adjacent Sertoli cell. (b) The germinal epithelium at lower magnification. Different stages of spermatogenesis in different regions of the seminiferous tubule, which might be seen in a single microscopic section of testicular tissue. Labels: a = Sertoli cell; b = spermatogonium; c = spermatocyte (easiest ones to see –pachytene stage); d = spermatocyte at a different stage; e = spermatid; f = Leydig cells in intertubular compartment; g = sperms hanging onto a Sertoli cell. (c) The 14 stages of a spermatogenic cycle in the testis of the rat (based on Leblond and Clermont, 1952). A = Type A spermatogonia;

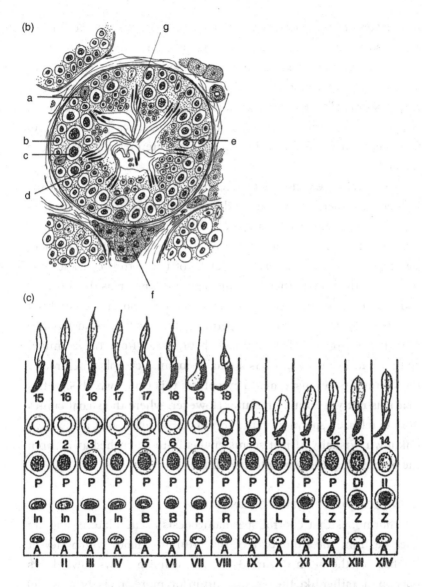

Caption for Figure 3.3 (cont.) B = Type B spermatogonia; In = intermediate forms of spermatogonia; R = resting spermatogonia; L, Z, P, Di = different stages of primary spermatocyte division (P is the easiest stage to recognize); II = secondary spermatocyte (sometimes not visible); 1–19 = stages in the morphogenesis (development of definitive structural features) of a sperm within the germinal epithelium.

to the complexity, the wave does not seem to exist in the human testis, so in man, more than one stage of spermatogenesis may be seen in a single section of tubule. Even in animals, the stages are not always sequential along the tubule, and when they are, they may progress towards the periphery (outside) or towards the centre of the testis. Staging spermatogenesis, therefore, requires quite a bit of practice, even when the basic principles of interpretation have been learned.

It will be seen later that the Sertoli cells exert a vital controlling influence on spermatogenesis, and the fact that elongated spermatids become deeply embedded into them is interesting. At this stage the sperms still have a remnant of Sertoi cell cytoplasm attached to them (see Figure 3.3a), but when they let go of the Sertoli cell and pass freely into the lumen, they lose this remnant (known as the 'residual body') and it is swallowed up by the Sertoli cell and thus taken back into its cytoplasm. It is as if the sperm needs some kind of sustenance from the Sertoli cell before finally being freed from the germinal epithelium. Even then, it has a small proximal cytoplasmic droplet just behind its head (spermatid cytoplasm this time), which signifies that it is not yet entirely mature. However, where the signal comes from to release a sperm from the Sertoli cell, or even what it is, is still unknown. Visually, it just seems gradually to be pushed away from the germinal epithelium into the lumen.

CONTROL OF SPERMATOGENESIS

The hormones that control spermatogenesis accumulate in and are released from the front part of the pituitary gland, which lies at the base of the brain (Figure 3.4a). The gland narrows towards its upper part, rather like the end of a turnip (or, more precisely, a swede) to form a stalk. This connects to the basal area of the brain known as the *hypothalamus*, so named because it lies beneath another part of the brain called the thalamus. It is, of course, full of nerve fibres, including neurosecretory fibres, which can exude hormones generated within the organ.

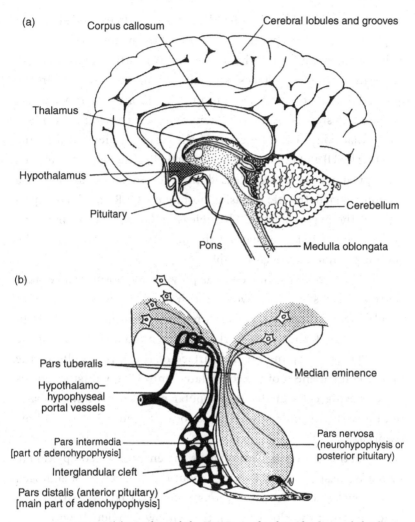

(a)

Corpus callosum

Cerebral lobules and grooves

Thalamus

Hypothalamus

Pituitary

Pons

Cerebellum

Medulla oblongata

(b)

Pars tuberalis

Hypothalamo–
hypophyseal
portal vessels

Pars intermedia
[part of adenohypophysis]

Interglandular cleft

Pars distalis (anterior pituitary)
[main part of adenohypophysis]

Median eminence

Pars nervosa
(neurohypophysis or
posterior pituitary)

FIGURE 3.4 (a) Locality of the pituitary gland in the base of the brain. (b) Portal system between the hypothalamus and the front part of the pituitary (anterior pituitary or adenohypophysis). Drawn by Ann Johnson.

Communication between the hypothalamus and the front (anterior) part of the pituitary gland is not solely through nerves, but finally by blood vessels forming a portal system. A portal system is one where blood is passed from one capillary bed (a network of the smallest blood vessels) to another, without entering the main circulation. A large capillary bed lies in a slightly expanded area of the pituitary stalk at a point where many of the neurosecretory fibres of the hypothalamus end (the area is known as the pars tuberalis). This tells us that substances, including hormones produced in the hypothalamus, ultimately pass into the anterior pituitary gland via this portal system. It is an important feature, because the two hormones that control the function of the testes, follicle stimulating hormone (FSH) and luteinizing hormone (LH), although accumulated in the pituitary gland, are released into the general circulation by a hormone (gonadotrophic releasing hormone – GnRH), which is produced in the hypothalamus. This releasing hormone can, therefore, only operate if it is transported through the portal system to the anterior pituitary (see Figure 3.4b).

FSH and LH are referred to as polypeptide hormones because of their basic structure, even though their precise structure is not identical. When they are both released into the general circulation by GnRH, they each work on different compartments of the testis. Special receptors on the surface of the Sertoli cells in the tubular compartment of the testis latch onto FSH, whilst receptors on the Leydig cells of the intertubular compartment specifically link up with LH molecules. Take a look at Figure 3.5 to see how it works.

FSH then enters the Sertoli cells and ensures the completion of spermatogenesis, whilst LH brings about the production of androgen by the Leydig cells. Almost everyone knows, I believe, that when androgens enter the bloodstream, they endow individuals with their sexual drive (that is, a compulsive desire to mate), but testosterone also passes directly into the tubular compartment of the testis by seeping through its basement membrane.

The outside rim or periphery of a seminiferous tubule is interesting, because outside the membrane is a layer of cells (the myoid layer – see Figure 3.3a), between which there may be more tight junctions, but periodically narrow channels. Small molecular-size substances such as steroids are able to pass through. This layer, by forming a

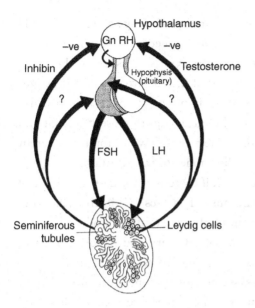

FIGURE 3.5 Illustration of the remote control of the testis by the pituitary. ? = possibly Taken from Glover (1988). Drawn by Ann Johnson.

selectively porous barrier, assists the Sertoli cell junctions to discriminate between those substances that are permitted to have access to the spermatogenic cells and those which are not. Proteins with large molecules, for example, do not have access, but smaller molecular-size substances such as testosterone have right of passage. This discriminatory barrier is known as the 'blood–testis barrier' and is similar to the 'blood–brain barrier'; that is, each acts as a sort of sieve between blood and tissue fluid, one being for testis fluid and the other for cerebrospinal fluid.

Once testosterone emerges through the barriers, the Sertoli cells ensnare it (they have an androgen-binding protein on their surface) and make it available to dividing spermatogenic cells. It is thus apparent that male sex hormone is also needed for the actual production of sperms. Also clear is the reality that if there were to be no curb on the production of testosterone, the consequences could be catastrophic. Just imagine what might happen if it got out of hand. Not surprisingly, therefore, there is a mechanism of control. When the level of testosterone in the bloodstream reaches a certain point, it tells the hypothalamus – and probably the front part of the pituitary also – to back off or slow down as far as producing and releasing FSH and LH are concerned. This is known as a 'feed back' mechanism, and it is interesting that testosterone conveys its message, to the hypothalamus at least, by being converted into oestrogen once it arrives there! This is less surprising than it might at first seem, because oestrogens and androgens are chemically quite closely related and are both known as 'steroids'. They thus have a totally different structure from the polypeptide hormones of the pituitary.

It would be reasonable to ask if there is a similar sort of control of Sertoli cell functions. The answer is 'yes'. This time, another hormone (not a steroid, but a protein) called inhibin, produced within the Sertoli cells, feeds back to the pituitary and hypothalamus to control the level of FSH (this is also demonstrated in Figure 3.5).

In spite of controlling systems, testosterone levels in the circulation are not constant, but episodic. There are regular high-level peaks or spikes and low-level troughs throughout a day, and overall higher levels at certain times of the day. A single estimate of testosterone, therefore, tells us virtually nothing about actual activity of the hormone in a given individual. Yet some clinicians continue to order one-off tests for testosterone. They could be meaningless!

We usually think of serotonin, which is a precursor of melatonin, as being hypothalamic, but it is also to be found in the testis. However, its role there is not yet absolutely clear, although it does seem to have an inhibitory effect on testosterone production and causes constriction in the testicular artery to an extent that it can render the testis virtually devoid of blood.

SPERMATOZOA: PRODUCT OF THE TESTIS

When sperms let go or are released from the Sertoli cells into the lumen of the seminiferous tubule, they first pass into a region of the testis where all the tubules converge to form a small reservoir (*rete testis*). They then pass into and through a series of tiny ducts (*ductuli efferentes* or efferent ductules) to enter the main excurrent duct of the testis, which is called the *epididymis* (Figure 3.1b).

The number of seminiferous tubules emptying into the rete testis varies in different mammalian species, but however many there are, they fan out throughout the testis and lie adjacent to each other. Interstitial tissue between them makes the testis rather like an orange with lots of very thin segments. In some species the testis consists of distinct lobules, so that clusters of tubules are contained by a sheet of connective tissue.

Sperms do not move at all or hardly at all within the seminiferous tubules or male reproductive tract. They are not motile here, because there is very little oxygen available. However, when they are exposed to oxygen and to the seminal plasma (as when they are ejaculated) they become, to use the jargon, vigorously motile. In other words, they start dashing about all over the place. This is because they have tails and can swim. Together with other special features, the ability to swim makes sperms unique among mammalian cells. In essence, they are lumps of DNA with a tail on them and with mechanisms for egg penetration. The typical structure of a mammalian sperm is shown in Figure 3.6.

First, unlike their precursor cells in the testis and other cells in the body, ejaculated sperms are quite independent and have virtually no cytoplasm (the protoplasmic substance that surrounds the nucleus of other cells). The nucleus of sperms is represented by the head, but the DNA which it contains is homogeneous, which is to say that it is so tightly packed that individual chromosomes, typically visible microscopically in the nucleus of other cells, cannot be seen in sperm heads. The front portion of the sperm head is covered by a double-layered cap, rather like a two-layered egg cosy. This is called the *acrosome*. It contains enzymes that, when released from the sperm, aid in fertilization by enabling sperms to penetrate the outer envelopes of the egg (see Figure 3.7). The process of spermatogenesis in the testis is a complex affair and we now know that there is quite a lot of plasticity (ability to change form) in the spermatogenic cells, largely due to a good deal of passage of substances in and out of their nuclei.

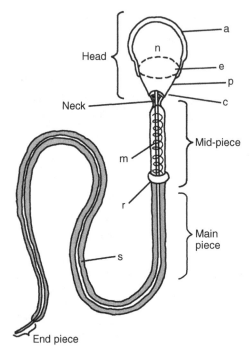

FIGURE 3.6 Structural features of a mammalian sperm. Labels: a = acrosome; e = equatorial segment (shadow caused by thinning of the lower part of the acrosome); p = post-nuclear cap; c = origin of axoneme (from the elongation of an organelle in the spermatid called a centriole – it lies just beneath a head plate at the distal end of the sperm head); m = spiral of mitochondria in the mid-piece; r = ring centriole (another centriole originating in the spermatid); s = protein sheath round the main piece.

Although, during spermatogenesis, the chromatin in sperm heads is finally rendered transcriptionally (functionally) inert by substances known as histones, and is tightly packed, some differential packaging occurs and some of the chromatin might be less inert than the rest. It is obviously beneficial for most sperm chromatin to be inert, because the sperm has a long way to travel before it can fertilize an oocyte, but this differential packaging is currently being studied in depth to determine exactly what is going on. Also, RNA (ribose nucleic acid or ribonucleic acid – a nucleic acid through which information from DNA is passed to produce specific proteins) has been identified in sperm heads and its presence there is interesting. It could be a left over from the process of the shutting down of DNA activity in spermatid and sperm nuclei, it could play a part in protein turnover in the sperm head during sperm maturation in both male and female tracts, or it might be involved in gene expression after fertilization has occurred. We do not yet know and research into

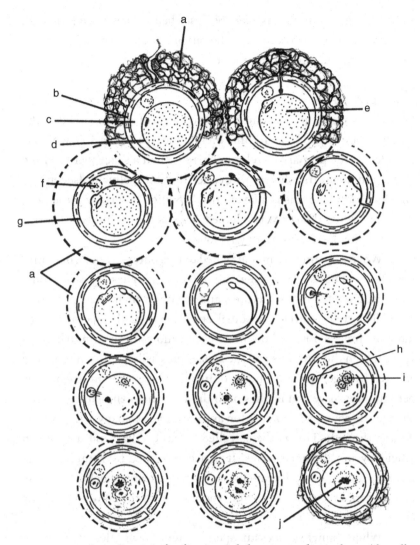

FIGURE 3.7 Events in fertilization. Labels: a = cumulus oophorus (the cells surround the whole zona pellucida and are simply cut off in this series of diagrams for reasons of space. They remain in reduced and flattened form long after fertilization and even up to the stage of early embryonic development (see last diagram); b = zona pellucida (semi-opaque); c = perivitelline space; d = vitelline membrane; e = vitellus (cytoplasm of the egg); f = first polar body (see text); g = zona changes its chemistry to keep out other sperms (zona reaction) after first sperm goes through; h = extrusion of second polar body as sperm enters the vitellus; i = male and female pro nuclei come together and fuse; j = the outcome of the fusion, the nucleus of the zygote. Modified from Austin (1965).

determining this might just now be described as 'a work in progress'. However, it is the acrosome and the tail of a sperm, together, that get a sperm into the vitellus in order to fertilize.

FERTILIZATION

Eggs or oocytes (the substance or cytoplasm of which is called the 'vitellus') are bound by a vitelline membrane (see Figure 3.7), which is regarded as the primary envelope. There is then a space (perivitelline space) and then a translucent ring called the zona pellucida (secondary envelope). Outside this is a layer of cells, which originally came from the ovary when the egg was released. This is the *cumulus oophorus*, which is the outer or tertiary envelope.

When a sperm becomes attached to this outer envelope or cumulus, fusion of parts of the inner and outer membranes of its acrosome form little bubble-like structures, and the outer membrane soon ruptures. This releases the acrosomal enzymes from the sperm surface (now the inner acrosomal membrane) and allows the sperm to bore its way through the cumulus and the zona more easily. In this, it is greatly assisted by vigorous movements of its tail. Once a sperm reaches the perivitelline space and latches onto the vitelline membrane, the zona changes its chemical composition and prevents too many other sperms getting through. This is the so-called 'zona reaction' and it varies in efficiency in different species. It is accomplished by granules being released from the vitelline membrane when a sperm touches it. These granules traverse the perivitelline space to reach the zona and thereby change its composition.

When a single successful sperm enters the vitellus by fusion of male and female membranes (thus carrying male DNA into the egg), the vitelline membrane forms a back-up block to make absolutely sure that no further sperms enter. This is known as the 'vitelline block to polyspermy'.

The rupture of the sperm surface, which is the starting signal for penetration of the outer envelopes of the egg by enzymic release, has been described in detail and is referred to as the *acrosome reaction.* The process of sperm penetration of an egg is demonstrated in Figure 3.7, in which it is shown that the sperm head and its tail separate once the sperm is in the vitellus. Incidentally, sperms that are not fortunate enough to be able to penetrate an egg pass along the Fallopian tube and enter the abdominal cavity, where they eventually die and disappear.

Another microscopic body seen in Figure 3.7 is the *polar body.* This is not really relevant to this discussion, but probably requires definition. The production of eggs is similar in principle to spermatogenesis, but mitosis of an oogonium in an ovary yields a primary oocyte and the first polar body (rather than two oocytes), and then meiosis produces a secondary oocyte and a second polar body. Both polar bodies (the second one being haploid) are thrown out of the oocyte into the perivitelline space, and the meiotic division is brought about by the entry of the sperm into the vitellus. This is a process known as *activation* of the oocyte.

Thus, sperms penetrate a secondary oocyte, so that strictly speaking, an egg (ovum) is only produced after extrusion of the second polar body. But in bitches, it is the primary oocyte that is penetrated by a sperm. However, we use the term 'egg' rather loosely and often refer to eggs when we really mean oocytes.

The part of the sperm other than the head consists of its flagellum or tail. This is attached to the head by small strands of tissue, which form the 'neck' of the sperm (see Figure 3.6). It is a fragile attachment and heads and tails are easily separated (decapitation). In fact, when a sperm is artificially inserted into a human oocyte (egg) (intracytoplasmic sperm injection – ICSI), the head and tail of the sperm are deliberately separated beforehand, so that only the head with its contained DNA is inserted. This shows that the tail is not necessary in fertilization itself, but only in enabling the sperm to get to the right place! Indeed, in amphibians, it never enters the egg at all, because decapitation occurs before the sperm enters the vitellus. In mammals, however, we have seen that normally (Chinese hamsters usually being an exception) separation of head and tail occurs within the vitellus after the sperm has entered the egg. The sperm tail then just appears to dissolve and disappear.

THE STRUCTURE OF SPERM TAILS

The first part of the sperm tail is rigid and contains mitochondria (small bodies usually found in the cytoplasm of cells) arranged in spiral form. These are the structures that generate energy for the tail to lash about; they are derived from mitochondria within the cytoplasm of the spermatid. This region is called the *mid-piece* of the sperm and is its 'power source' or 'engine room'. The rest of the tail is less rigid and is capable of a series of bending waves, which give rise to a type of figure-of-eight pattern of movement towards its end. This area of the tail is known as the *main piece* and is surrounded by a protein sheath. The very end of the tail is not covered by this sheath and is called the *end piece*. The whole sperm is covered by a typical plasma (cell) membrane.

Before the definitive structure of an adult sperm, as shown in Figure 3.6, can be achieved, several maturation changes need to occur after it is first released from the germinal epithelium. In some species there are changes in the acrosome during maturation of the sperm, and the droplet of spermatid cytoplasm moves along from the neck of the sperm to the end of the mid-piece, so that it is easily lost during ejaculation. These maturation changes in structure occur as the sperms traverse the main excurrent duct of the testis (the epididymis). It is important to understand that sperm tails have a peculiarly complex architecture and are certainly not like a simple whiplash.

Sperm tails have a core or *axoneme*, as shown in Figure 3.8. This consists of two central microtubules (small microscopic tubules) and it is surrounded by nine outer microtubules, appearing as couplets or doublets. The inner microtubule of each doublet has two little arms on it, called 'dynein arms' (an outer one and an inner one), so called because they consist of a protein, dynein. These doublet microtubules are joined together by another protein called nexin and they are surrounded by nine much larger tubules which lend a

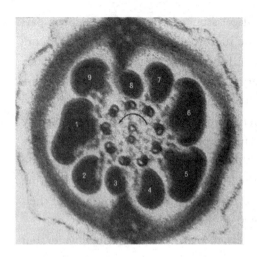

FIGURE 3.8 Electron micrograph of a section of a sperm tail, showing nine inner doublet microtubules and nine outer large asymmetrically placed microtubules enclosed in the protein sheath (see text). Note the dynein arms on the inner microtubules. The anticlockwise arrangement of the outer tubules means we are looking up the flagellum from the tip to the base. From Fawcett (1966).

certain rigidity to the upper end of the tail. However, some of these larger microtubules drop out towards the end of the tail, allowing the tail to move more freely. For this structure, refer again to Figure 3.8. When we get down to real detail, the microtubules consist of bundles of subfibrils.

So, how does the energy produced by the mid-piece mitochondria reach the rest of the tail? An enzyme created in these mitochondria (ATPase) diffuses down the length of the flagellum and enables the energy needed for movement to be released. But such movement is dependent on the presence of two or three proteins, which react with one another in order to elicit contraction in the tiny subfibrils within the microtubules. One of these proteins is dynein (another, for example, is tubulin), so dynein is essential for flagellar movement. This has been clearly demonstrated in humans by a condition known as Kartagener's syndrome, in which dynein arms are absent. The sperms of these patients are completely incapable of movement and they just lie around statically in the ejaculate.

But why do sperms swim at all? They do not swim through the female tract, but are carried to the site of fertilization by outside forces. We can speculate that the reason sperms have to swim is because otherwise they might get stuck in the cervix, the lower part of the Fallopian tube or in the uterotubal junction, all of which are rather narrow areas of labyrinthine structure. Supporting this idea is evidence that if dead sperms (obviously with no movement) are placed in the vagina, they will not enter the uterus. Moreover, we also know that flagellar movement is needed, in spite of the acrosome reaction, to help propel sperms through the outer envelopes of the egg. It is important to stress, therefore, that the idea of mammalian sperms swimming through the female tract to the site of fertilization is a myth. They swim in all directions and the only purpose of their tails is to make sure they keep on the move.

When fertilization is external, sperms swim through water, but when it is internal, they simply swim in random directions in the seminal plasma or in the milieu provided by the female tract, and they are carried to their destination largely by waves of uterine contraction. But all sperms are not flagellate anyway. Those of crabs, just as one example, are star-shaped. It all depends on the evolution of sexual circumstance.

SPECIES DIFFERENCES IN THE APPEARANCE OF MAMMALIAN SPERMS

The detailed appearance of mammalian sperms is fairly characteristic of the species that produces them, and species differences thus exist, for example, in the length of sperm tails and particularly of the mid-pieces. Species differences in the sperm nucleus and especially the acrosome are most obvious.

These are especially noticeable in the shape of sperm heads. Although there is evidently little difference in the length of sperm heads between species, some, such as those of a boar or bull, appear, subjectively, to be quite large when viewed microscopically. The same applies to those of some of the antelopes. But bovine animals seem to have especially large sperm heads, as strikingly exemplified by those of the eland bull. This is in marked contrast to the heads of elephant sperms which, in relation to the size of the animal, appear comparatively small. Amazingly, the head of guinea pig sperms looks as big as, if not bigger than, those of many larger animals. These differences cannot reflect quantitative differences in the DNA content of the nucleus (though differences in the total complement of chromosomes occur in different species), but must rather depend on the tightness of DNA packing and the shape of the head, remembering that microscopically (using the light microscope) a sperm is being viewed only in two dimensions. Some sperms are globular, whilst others are flatter, so a false image of differences in size might be obtained.

Variations in the shape of the acrosome are more bizarre. In larger animals, the acrosome fits snugly around the nucleus, but in guinea pigs it is more expanded (this probably makes the head look bigger than it is). Typically, the sperms of small rodents have hooked acrosomes. These include those of rats and mice and a host of wild rodent species.

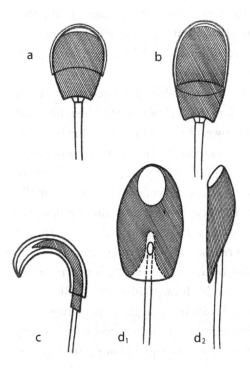

FIGURE 3.9 Comparative diagrams of sperm heads (not drawn to scale) showing variations in the size and structure of acrosomes. (a) Rabbit. (b) Bull – typical type of sperm head structure in large animals, including stallions, all ruminants, boars and elephants, etc. (c) Hooked acrosome of rat. (d_1) and (d_2) Two dimensions of possum sperm. Drawn by Doug Bailey.

The sperm heads and acrosomes of the monotremes (prototheria) such as the platypus are unique, and the whole head can be seen to tilt on the end of the flagellum. Attachment between the sperm tail and the sperm head in some marsupials can also be fairly specific.

Even though there is an acrosome reaction, the hooked acrosome of rodents enables their sperms to penetrate the cumulus cells of an oocyte by a sort of hacking motion, whereas when there is no hook, the sperm head moves from side to side as it wends its way through the cells. Diagrams of some of these differences in sperm head and acrosomal structure are shown in Figure 3.9 and their photomicroscopic appearance is shown in Figure 3.10.

SOURCES OF SEMINAL PLASMA

The liquid portion of the semen – the seminal plasma – is produced by several *accessory organs* of reproduction, which include seminal vesicles (seminal glands), a prostate gland and bulbourethral glands (these are sometimes called Cowper's glands and lie close to the root of the penis). There are some glands that line the passageway to the exterior (the *urethra*); these are disseminated along the lining of the penis itself.

The epididymis may also be regarded as an accessory gland, because it produces fluid. The duct is long and coiled and lies adjacent to the testis, but it straightens towards its terminal part and becomes the ductus deferens or vas deferens. Finally, the vas deferens becomes expanded within the pelvic cavity to form an ampulla, and this probably contributes most of the total excurrent duct fluid. The ampulla enters the urethra alongside the ducts leading from the seminal vesicle, although in man and some other mammals these ducts join together before entering the urethra to form two ejaculatory ducts. This part of the urethra, being in the pelvic cavity, is referred to as the 'pelvic urethra' and it extends through into the penis as the 'penile urethra'. The prostate and bulbourethral glands also empty into the pelvic urethra (sometimes referred to as the 'prostatic urethra'). The whole arrangement of these glands is depicted in Figure 3.11.

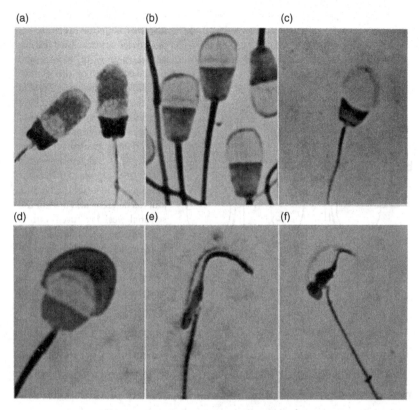

FIGURE 3.10 Photomicrographs showing variations in sperm head morphology (variable magnification, so sizes as they appear are not comparable). These particular sperms are stained with silver to show up the postnuclear cap, but well illustrate the different head shapes. (a) Boar sperms. (b) Bull sperms (original picture by J.L. Hancock). (c) Rabbit sperm. (d) Guinea pig sperm. (e) Rat sperm. (f) Mouse sperm. Reproduced from Walton (1968).

THE EPIDIDYMIS

The epididymis is an extremely interesting accessory organ in that it not only contributes, albeit in a small way, to the seminal plasma, but it also transports sperms away from the testis and stores them when they have become mature.

Biologists and clinicians refer to the epididymis as having a head (the upper part), a body (middle part) and a tail (terminal segment),

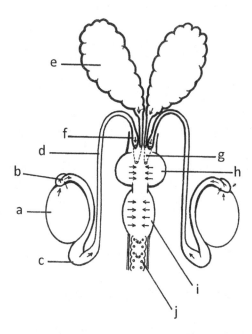

FIGURE 3.11 The male reproductive tract, showing the testes
and accessory organs and the direction of flow of sperms and
fluid secretions of the glands.
Labels: a = testis; b = head of
the epididymis (first part after
the initial segment); c = tail
of the epididymis (terminal
part); d = ductus (vas) deferens;
e = seminal vesicles or seminal
glands; f = ampulla of the ductus
deferens; g = ejaculatory duct
(ductus deferens and duct of the
seminal vesicles fused into a
single duct, such as in man, stallions and some other species);
h = prostate gland; i = bulbourethral (Cowper's) glands; j =
urethral glands (of Littre or Mor-
gagni's glands in man).

although more finely tuned and detailed descriptions of its structure
have been given. Most of the sperms probably mature during their
passage through the head region and upper body of the epididymis
(although some might be a little laggardly and complete their maturation lower down the duct), before being flushed through the body of the
organ into the sperm store, which is in the tail of the epididymis. There
has been some confusion, because accumulation of mature sperms in
the lower part of the body of the epididymis has led to the view that
sperms do not complete their maturation until they reach here. Whilst
it is possible some sperms may take all this time to mature, I think the
overall concept might not be correct and that most of them mature
higher up in the duct and simply accumulate at this level. In any event,
when these mature sperms reach the tail of the epididymis they are
stored until they are required at the time of ejaculation.

When a male individual is sexually inactive – and this can even be
in the mating season if there are no females around – the store can
overflow because of a *vis a tergo* caused by the continuous production

of sperms. On these occasions, sperms spill over into the vas deferens and may be voided in the urine. Also, if sexual inactivity is prolonged, some superfluous, ageing and disintegrating sperms are absorbed within the epididymis. This may be observed particularly easily at the end of the mating season, when androgen levels (which normally maintain epididymal function) start to decline. But species differ in the ability of their epididymis to absorb disintegrated sperms. The mechanism is most easily observed in the epididymis of old men and of older animals such as the rat and hamster. The rabbit epididymis is not very good at absorbing sperms, although it can do so. Presumably the mechanisms for sperm absorption in the rabbit epididymis are not well developed, but rabbits are rarely sexually quiescent anyway, so there may not be a need!

So, there are mechanisms within the epididymis for the destruction, dissolution and absorption of sperms under certain circumstances. So why are some sperms destroyed during normal epididymal passage and others not? Obviously, some are more vulnerable and less robust than others, but it must be that for the majority, survival mechanisms prevail over destruction mechanisms. In unusual conditions such as the withdrawal of androgen or increased temperature, this situation must be reversed, because all contained sperms soon die off under these conditions.

The rate of flow of sperms through the epididymis is not constant either, due to fluid being absorbed or secreted at different levels of the duct (Figure 3.12). Fluid produced in the testis by the Sertoli cells is absorbed by the efferent ductules (ductuli efferentes) leading from the testis into the epididymis and then also in the very first part of the epididymis, called the 'initial segment' (see also Figure 3.12). This results in a tendency for sperms to be seen as being tightly packed in the efferent ductules, but they shoot through the initial segment pretty quickly (due to extra active muscle in this region), so very few of the concentrated sperms are usually to be seen in the lumen here, and as they pass into the head of the epididymis, more fluid is quickly added to dilute them (so they immediately appear to be diluted in the first part of the head of the epididymis), only for the

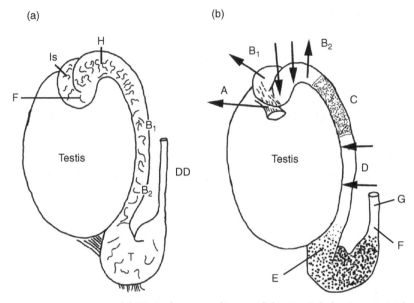

FIGURE 3.12 (a) Typical mammalian epididymis. Labels: Is = initial segment of the epididymis (it is into this part that the efferent ductules empty); F = the epididymis turns back on itself to mark the head of the epididymis (caput flexure); H = head of the epididymis; B_1 and B_2 = upper and lower body of the epididymis; T = tail of the epididymis (terminal segment); DD = ductus deferens (vas). (b) Areas of fluid absorption and secretion in typical ruminant epididymis. Labels: A = efferent ductules; B_1 = initial segment (really part of the head); B_2 = head of the epididymis proper; C = last part of the head and first part of the body, where sperms become very concentrated and probably represents the main area of sperm maturation; D = main body of the epididymis (fluid added and sperms flushed into the tail of the epididymis (sperm store)); E = tail of the epididymis (sperms fairly concentrated, not due to fluid absorption so much as simple accumulation and slowed flow; this is caused by F = duct narrows; G = ductus deferens (normally contains just overspill of sperms).

fluid to be reabsorbed a little further along the duct to concentrate them again. This secretion and resorption of fluid can only result in an overall slowing of the passage of sperms, giving them adequate exposure to maturation antigens in the epididymal milieu. Next, more fluid is added in the upper part of the body of the epididymis to carry sperms (most of which should, by this time, be mature) into the sperm store. Much of this fluid, plus ampullary fluid, is discharged at ejaculation, together with the other secretory products.

The multiple functions of the epididymis require it to be a long tube and to be extremely coiled, so as to fit neatly into the small space offered by the scrotum. It has been recorded as being 70 metres long in a stallion, 40 metres in a bull and 6–7 metres in a man. The sperms, therefore, take several days to traverse the duct before reaching a point towards the end of the sperm store, where they can be ejaculated. They are passed through the epididymis by regular contractions of the duct (peristaltic contractions, rather like those of the gut).

Embryologically, the epididymis started out as the mesonephric duct (sometimes referred to as the Wolffian duct, but this is a questionable appellation, since it tells us nothing about its origins or function). It is thus evident that it is originally the duct of a primitive kidney (mesonephros) in the embryo and as development proceeds, it is hijacked by the testis and retained as its own. This doesn't matter, because the embryo soon develops a more sophisticated kidney (metanephros) with its own duct, but it is not surprising (since it is essentially a kidney duct) that the epididymis displays selective secretion and reabsorption just as kidney ducts and tubules typically do. It just happens to be very useful in accomplishing sperm maturation.

CONSTITUENTS OF SEMINAL PLASMA

The significance of seminal plasma is difficult to assess. It contains a variety of chemical substances, including the seminal sugar, fructose. When mixed with sperms, seminal fructose gives a boost to their motility and, in most mammals, it is largely produced in the seminal vesicles. Several other substances appear in these glands, and some of them may be of assistance to sperm survival. Yet other substances are yielded by the prostate and bulbourethral glands. It is fair to say, though, that sperms do not like seminal plasma very much, and after being ejaculated into the female tract, they do their best to get out of it as soon as possible. For example, even mature sperms undergo a further process of development when they get into the female tract. This is called *capacitation* (a sort of preparation for the acrosome reaction) and the seminal plasma has in it a decapacitation factor, so if sperms stay in the fluid too long, they are unable to capacitate. Further exploration of this situation could be useful, because semen may contain some intriguing substances such as inositol, ergothioneine and carnitine and, in some species, especially some non-mammalian vertebrates, serotonin. These are also to be found in the male tract, along with other substances such as glyceryl-phosphorylcholine in the epididymis.

All sorts of materials in the bloodstream can make their way into the semen through being secreted by the seminal vesicle. The seminal vesicle can act as a kind of dustbin, and to an extent reflects what is going on in the blood. An individual with jaundice, for instance, ejaculates yellow semen (if he feels well enough to ejaculate at all), because the bile pigments, which normally pass into the gut from the gall bladder, dam back into the bloodstream because of obstruction in the bile duct. Doubtless, various dyes and other contents of the blood could also be passed into seminal fluid via this route.

However, although seminal plasma may not be very popular with sperms, it can exert an effect on the female tract. It contains substances called *prostaglandins*, which, when absorbed through the wall of the female tract, cause increased contractions of the tract, which aid the transport of sperms to the site of fertilization, probably through facilitating the action of oxytocin, which is known to increase uterine contractions. Prostaglandins may influence the ovary as well, incidentally.

SPECIES DIFFERENCES IN SEMEN

Broadly, the quality of semen is specific to each species of mammal and, not surprisingly, depends on the structure and development of the testes and accessory organs in each case. Having explored the basic theme of semen production, it is now appropriate to examine the relationship between the nature of ejaculated semen and the organs that produce it, as it applies to different species. Differences in the arrangement of accessory organs are illustrated in Figures 3.13–3.17.

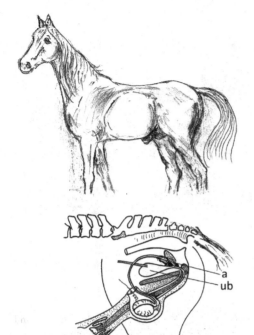

FIGURE 3.13 Tract of a stallion showing the position of the accessory organs. Labels: a = large ampulla of each ductus deferens; ub = urinary bladder.

If we take animals with a large semen volume first, we find that a stallion's ejaculate is, on average, 70 ml (varying from 30 ml to 300 ml) and that of a boar is usually described as being 250 ml, but it may reach as much as 600 ml and can even almost fill a milk bottle. In each of these cases, the seminal vesicles or seminal glands are large, being glandular in a boar and appearing as large sacks in a stallion. Bulbourethral glands in a boar are large, and the ampullae of each ductus deferens in stallions are huge. On the basis of these structural features, we would rightly anticipate that the volume of fluid these animals produce would be large, so it is not surprising that it is. Similar features apply to other equine species such as the ass and zebra. In the same way, wild boars and warthogs broadly share the same features as the domestic boar. Dogs typically produce 7–9 ml of semen, but this results entirely from a large and highly developed prostate. Dogs have no seminal vesicles or bulbourethral glands and therefore have no fructose in their semen.

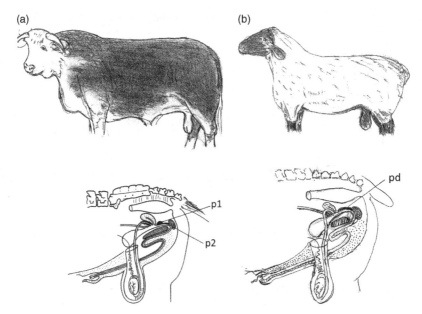

FIGURE 3.14 Farm animals. (a) Tract of a bull. Pendulous testes and divided prostate. Labels: p1 = compact prostate; p2 = disseminated prostate. (b) Tract of a ram. Pendulous testes. Disseminated prostate only. Label: pd = disseminated prostate.

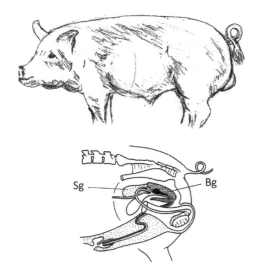

FIGURE 3.15 Farm animal. Tract of a boar. Very large seminal vesicles (glands) and large bulbourethral glands. Labels: Sg = seminal glands; Bg = bulbourethral glands.

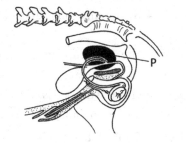

FIGURE 3.16 Small animal. Tract of a dog. No seminal vesicles or bulbourethrals. P = large prostate.

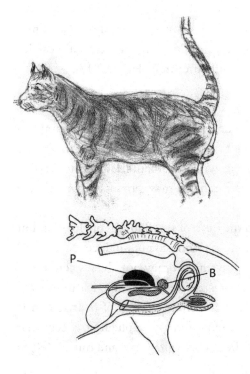

FIGURE 3.17 Small animal. Tract of a cat. No seminal vesicles. Labels: P = prostate; B = very small bulbourethral glands. Note also, backward pointing penis when not erect.

As might be expected from the above, animals with small volumes of ejaculate have less well-developed accessory glands. In bulls, for example, although the seminal vesicles can be felt through the rectal wall, they are not especially big and they have no large cavity for storing fluid, like those of a stallion. The vesicles of the bull are not really vesicles at all, in that they are almost entirely glandular – hence the term 'seminal glands'. This also applies to the vesicles of sheep and goats and to wild bovine species and antelopes. Bulls have an ejaculate volume of only 4–5 ml; rams and goats 1–2 ml; and wild ruminants are similar.

There is also some variation in the structure of prostates. In a bull, for example, the gland is in two parts. There is a compact part and another part where the tissue is disseminated along the wall of the pelvic urethra, whilst in rams it is purely disseminated and is thus devoid of a compact part (Figures 3.14a and 3.14b).

Semen volume in domestic cats is very small (less than 1 ml); this is also the case among wild cats. Cats have no seminal vesicles either, but they do have small bulbourethrals (Figure 3.17).

SEMINAL GEL

The semen of larger animals such as stallions and boars contains a jelly-like substance called 'gel'. This is notably absent from the semen of ruminants, including wild ruminants, and from that of dogs and cats. But rodents and rabbits may produce quite a large amount of gel.

In stallions the gel is in the form of small lumps or globules, but it is not a constant feature. Some stallions do not produce it at all, and successive ejaculates of the same stallion may not always contain it. Similar globules of rather larger size are a constant feature of the semen of boars. If the rest of the seminal plasma of boar semen is poured off, the gel has the appearance of tapioca pudding, except that the lumps are a little larger. The gel is hygroscopic and thus, if left to stand, it grows in size.

The gel of rabbit semen and that of smaller rodents is not ejaculated in lumps, but rather as a single mass of gelatinous material. Also, while the gel in boars appears to be formed during ejaculation and the bulbourethrals make an important contribution to its formation, the gel of rabbits probably comes from a part of a sack-like seminal vesicle (glandula vesicularis) and is usually or often formed in this seminal vesicle-like structure before ejaculation occurs. Again, as in some stallions, rabbits do not always yield gel in their semen.

Human semen has a gelatinous plug, but it is of unique quality, in that it is rather more of an opaque clot than a gelatinous blob. Moreover, it liquefies in about 20 minutes if left standing or, presumably, after it has been deposited in the female tract after coitus. A similar clot due to seminal coagulation occurs in the semen of other primates too. Opinions differ about the function and biological significance of gel; the matter will be discussed later in more detail.

SPECIES DIFFERENCES IN SPERM CONCENTRATION

As far as domestic animals are concerned, those with large ejaculate volumes usually have a relatively low concentration of sperms. The concentration varies both within and between individuals even in the same species, so different figures are given in different texts. Those given here are based on my own experience of measurement.

The concentration of sperms in stallion semen is between 120 000 and 150 000 per microlitre (µl) (though it can be up to 300 000 per microlitre); in boar semen it is about the same, but most boars have around 250 000 sperms per microlitre of semen. Bull semen, with its relatively small semen volume, has a sperm concentration of 1 000 000 per microlitre and the semen of rams and goats may contain 4 000 000 sperms per microlitre. Dog semen has 300 000 per microlitre and rabbits can also produce 300 000 or even more per microlitre, although it is usually about 150 000 (Table 3.1).

Table 3.1 *Sperm numbers and concentrations*

Species	Average semen volume (ml)	Total number of sperms ejaculated ($\times 10^8$)	Concentration of sperms per microlitre of ejaculate
Stallion	70	84	120 000
Ass*	40	120	300 000
Boar	150	375	250 000
Bull	4.5	45	1 000 000
Buffalo+	2.5	15	600 000
Camel+	8	32	400 000
Ram	1.5	60	4 000 000
Goat	1	40	1 000 000
Red deer*	4	8	200 000
Reindeer*	0.5	2.3	460 000
Dog	8	24	300 000
Fox+	1.5 (average, but very variable)	1.05	70 000
Cat	>1	0.4	50 000
Rabbit	1.5	3	200 000 (very variable)
Bat+	0.5		
Human	2.5	2	80 000 (can be double this figure or less)
Chimpanzee		6	
Gorilla		0.6–0.7	

* Mainly based on personal measurements and data from Dott and Glover (1999).
+ Data from Mann and Lutwak-Mann (1981).

THE SIZE OF TESTES

Understandably, the testes of sexually quiescent animals are smaller (sometimes dramatically so, as we have seen in hyraxes) than when they are sexually active. This is because in sexual quiescence there is no testicular activity, whilst in the mating season it is full steam ahead for spermatogenesis, and the tubular compartment of the testis dilates and is packed with sperms. Also at this time, the Leydig cells in the intertubular compartment increase in size, may multiply and burst into action. Although the testicular tunic is fairly inelastic, it stretches as much as it can, but it is tough and will not rupture. The massive increase in the size of the sexually active hyrax testis – and also in hedgehogs – means the tunic must also grow, otherwise it would have to rupture. These are seasonal changes, but there are discrepancies between species in the size of active testes as well.

If we look at differences in overall body size between species and then at the size of testes and of the sperms they produce, we are faced with some apparent conundrums. Obviously, an animal the size of a rat cannot have testes as large as those of a camel, since a camel's testes are bigger than the entire body of a rat. Broadly, therefore, the larger the animal, the larger its testes will be (there are exceptions, such as the great gorilla, the body of which can be massive, but its testes are quite small). But paradoxically, there is an inverse relationship between testis size and body size. Thus, relative to body size, the testes of small rodents are larger than those of an elephant or even a whale. A striking example of this is the Australian honey possum, which has enormous testicles relative to its size.

Within this broad framework, however, a number of additional factors operate to influence testicular size. Rams have very large testicles, which are situated vertically in the scrotum and dangle between the hind legs. This applies also to bulls. The testicles of stallions are relatively small and are carried in the scrotum horizontally and closer to the abdominal wall. The large testicles of boars swing backwards in descent and end up lying upside down, as it were, just under the anus, but close to the body wall. Those of dogs and cats are slightly more pendulous, yet still protrude at the back end of the animal. Refer back to Figures 3.13–3.17 to observe these differences in domesticated mammals.

Evolutionary forces must have been involved in the differences, so let us try and guess at some of them. Rams are polygynous, with a large flock of ewes to serve, so large testes are necessary in order to cope with sperm demand. Such large pendulous testicles must hinder the animal from running away from danger, but doubtless a sheep's distant ancestors were mountain beasts, and with their skill in such terrain they were probably less exposed to predators than animals on the plains.

Goats are mountainous and have smaller testicles, but they do not have the demand for sperm production equal to that confronting rams. Bulls are descended from animals that few would dare to approach; the buffalo, for example, is one of the most ferocious and feared animals in Africa. There is little need for it to run – although young buffalos can be chased by lions, even the lion has respect for adult buffalos and would not dare a head-to-head confrontation. Moreover, buffalos can turn swiftly to force a confrontation. But lions usually attack from the rear, so even a buffalo cannot tarry too long in coitus.

An enraged wild boar is hardly easy prey either, and in any case, its portly figure and short legs would cause some trauma of the testicles, should they hang too low. Unfortunately, before porcine animals grow in toughness, their short legs can be a disadvantage and occasionally young warthogs can also fall prey to lions.

Wild horses and zebras are prey animals, and their main defence – apart from kicking – is to outstay their predators in running. They have therefore evolved as long-distance runners, a role in which pendulous testicles would prove a serious drawback. As such, equine testicles are tucked neatly out of the way. The testicles of antelopes, although lying vertically in the scrotum, are also much closer to the abdominal wall than those of rams or bulls. They do not really get in the way, therefore, when the animal is sprinting or weaving in his escape. Similarly, camels have vertically situated testicles, but they are no impediment in fast movement, because camels have such long legs. The same applies to the vertical testicles of the giraffe. The kangaroo carries his testicles in front of the penis rather than behind it, but he bounds through the air as his mode of travel, so perhaps he feels that if he does not carry his testicles up front, they might get left behind as he flies through the air with

FIGURE 3.18 Water buffalo of Africa. If confronted head on, he's quick, fierce and frightening!

such impressive speed! Clearly, this is a convenient evolutionary development.

Dogs and cats are predators, so for a different reason, they need to be able to run fast also. Thus, we see their testicles situated in a fairly suitable position. The same applies to man, whose hunting ancestors surely needed to be able to sprint through the forest or wherever when chasing their prey.

Rams exemplify *par excellence* the relationship of testicular size and sperm demand, but it is also well illustrated in some of the primates. The chimpanzee is promiscuous and mates frequently, so he will need plenty of sperms. Not surprisingly, therefore, his testicles are large. By comparison, the gorilla, who, although polygynous, has usually no more than two females with him at a time, has diminutive testicles. No creature would be stupid enough to challenge him anyway, so he can afford to be a lazy mater and have no need for large testicles. Relative to these two primate species, human testicles are intermediate in size.

These vast species differences in testicular size are all very well, but sperm sizes are all pretty similar. So does this mean that a ram, for example, produces millions more sperms than, say, a rabbit? Does a stallion produce massively more sperms than a dog? Surprisingly, perhaps, though the relationship is not linear, the broad answer seems to be 'yes'. The total content of sperms in any ejaculate is worth examining.

SPECIES DIFFERENCES IN THE TOTAL NUMBER OF SPERMS EJACULATED

Because there are several factors involved, it is dubious to assume that the number of sperms in each ejaculate is directly related to the number that are regularly produced by the testis. But it is also logical to conclude that if a large number of sperms is demanded in each ejaculate, a correspondingly large number must be produced and stored in the first place. By multiplying the sperm concentration (per microlitre or millilitre of semen) by the ejaculate volume, the total number of sperms in an ejaculate can be calculated, meaning the relationship between this and testicular size (sperm production) can be tested.

Bulls, rams and boars have the biggest testes among the larger domestic animals, so they would be expected to produce and ejaculate more sperms than others with smaller testes. If we take a concentration of sperms in boar semen as being, minimally, 150 000 000 per millilitre and consider that the animal produces such a large volume of semen (say, 250 ml), we can expect a yield in an ejaculate of around 37 500 000 000 (375×10^8) per ejaculate, and if the concentration is 250 000 000 per millilitre, the yield will be 625×10^8. Sperm concentration in rams is about 4 000 000 000 per millilitre, so each ejaculate will contain between 4 000 000 000 and 8 000 000 000 (40–80×10^8), according to the volume of any particular ejaculate. In bulls, a sperm concentration of 1 000 000 000 per millilitre will also yield 4 000 000 000–5 000 000 000 sperms (4–5×10^8) in each ejaculate. If a stallion gives 120 000 000 sperms per millilitre, then an ejaculate of 70 ml would contain 8 400 000 000 sperms (84×10^8).

A smaller mammal such as a dog has a sperm concentration of 300 000 000 per millilitre in an ejaculate of 8 ml in volume. Thus, each ejaculate will only contain 2 400 000 000 (24 × 10^8). We could estimate that cat semen will have about 40 000 000 (0.4 × 10^8) sperms in it and rabbit semen only 300 000 000 (3 × 10^8). So it seems that larger testes do indeed yield many more sperms in ejaculates than smaller ones. This is presumably why, during the course of evolution, the imbalance has been partly overcome by the development of an inverse relationship between body size and testicular size.

Humans are poor fish compared with the above species, because they ejaculate a mere 2–4 ml of semen (average 2.5 ml), and the sperm concentration is around 80 000 000 per millilitre. This means that, on average, a human male ejaculates little more than about 200 000 000 sperms in each ejaculate (2 × 10^8). Gorillas ejaculate about 60 000 000–65 000 000, or 0.6–0.65 × 10^8 sperms per ejaculate, but the promiscuous chimp, with his large testicles, discharges up to 650 000 000 (6.5 × 10^8) (Table 3.1). Thus, it seems that species with a promiscuous way of life, or a demanding polygynous one, ejaculate a lot of sperms, because coitus is frequent and they need to keep up the numbers (Table 3.1).

These data will normally be modified by frequency of mating. Horses and pigs do not copulate as often as most ruminants, for instance, and it might at first appear that they will need fewer sperms. But they produce very large quantities of seminal plasma, so any major restriction on sperm numbers would make each sample so dilute that it could fail altogether to bring about fertilization. So sperm numbers here are very much associated with semen volume. This is not the case with high sperm producers with small semen volumes such as rams, because all they are asked to do is to keep up the numbers in a scenario of frequent mating. This is very important, because frequent mating itself influences sperm concentration even within a species, and tiring of the emission reflex manifests itself in the delivery mechanism of sperms more markedly than in the volume of seminal plasma.

In rabbits it has been shown that sperm concentration halves in a second ejaculate if two are produced in quick succession. This decline is less apparent in rams (which is fortunate when we consider the number of ewes inseminated by the one ram in a day!), and in any event, semen volume is soon affected also, though this is less marked in rams. This helps sperm concentration to be maintained.

However, it is important for frequent copulators such as rams and chimpanzees to have a goodly number of sperms in their semen at the outset, just in case, after a hard day's work, they run out completely. Using hypothetical figures, if you start off with 1×10^8 sperms (100 million) and the number falls by 0.1×10^8 (10 million) with each successive ejaculate, you would theoretically run out after ten ejaculates, but if you started off with 2×10^8 (200 million), you would not run out until you had ejaculated 20 times. This is just simple arithmetic, but demonstrates that if a large number of viable sperms is required in a single day, as in some promiscuous or polygynous species, the more sperms that are present in the first ejaculate (as a rough indicator of output), the greater the safeguard will be against running out later on. Such species, therefore, usually have large testes.

However, the overall size of testes can vary within the same species. In cattle, this is often according to breed, as it may also be in rams. Suffolk rams have larger and firmer testes than some other breeds such as Romney Marsh or Hampshire Downs. In individual humans, large testes usually obey the rules and signify a high sperm output, but the testes of East Asian men are known to be smaller, on average, than those of Caucasians. Yet, surprisingly, there is no significant difference in sperm count between the races. It can only be that the tubular compartment of the testes of Oriental men is more convoluted and packed more tightly in the ball than that of Occidental men. Also, they may numerically have more seminiferous tubules. The same must apply to variations in the size of ovine testes.

So far the focus has been primarily on sperm production by the testes of different species, but, except in man, the number of sperms in successive ejaculates is also a reflection of the extent of sperm storage.

SPERM STORES

In the majority of mammals, sperms are stored in the tail of the epididymis, and in the past there has been so much concentration on sperm maturation as a function of this duct that its all-important function in sperm storage has tended to be overlooked.

Capacious sperm stores usually accompany large testes and may thus signify a substantial sperm demand as much as the testes themselves. A reservoir of sperms in the tail of the epididymis is produced by a partial damming back of sperms from the junction of the epididymis and the vas deferens. This is because there is a sudden narrowing of the passageway (lumen) of the duct at this point. Although there might be some spontaneous overflow of sperms into the vas deferens from the tail of the epididymis, when an animal is sexually quiescent but still actively producing sperms, it is slow, and only in seminal emission is a significant number pushed into and along the vas. This is covered in Chapter 5.

In rams, the high frequency of mating means not only that a lot of sperms need to be produced, but also that there should be a store of sufficient size to ensure that an adequate number of them can be held at the ready after they have matured. Thus, the tail of the epididymis in rams is expansive, and in the breeding season it is packed with sperms. This is to be seen in other seasonally breeding ruminants such as the deer (see Figures 3.19 and 3.20). To a lesser extent it also applies to bulls and, especially, boars, if for a rather different reason. Boars do not mate as often as rams, but it is to be remembered that they have an obligation to ensure there are enough sperms to occupy the large volume of each ejaculate. The same applies in some degree to stallions, which also have a fairly well developed sperm store.

The tail of the epididymis in promiscuous animals such as rabbits and rats is also quite large, as a response to frequent demand. This is also the case with promiscuous primates such as the chimpanzee, as well as the baboon, which may copulate up to 25–30 times in a day. It would be a famous human primate who could match that!

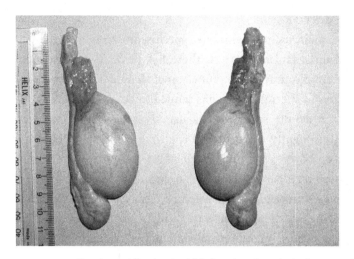

FIGURE 3.19 Ram's testicles (each side) showing the relatively extensive sperm store in the tail of the epididymis.

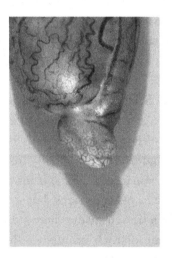

FIGURE 3.20 Large sperm store in the tail of the epididymis of another seasonally mating ruminant, the red deer.

But a large sperm store is not required in man, who, on average, copulates only 2–3 times per week. Accordingly, we see that the tail of the epididymis in man is very poorly developed (see the differences in Figure 3.21). If anything, the vas deferens and its ampulla do a better job of sperm storage in man; that is, if significant storage occurs at all. Similarly, the male nine-banded armadillo has a poorly

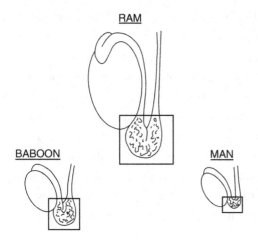

FIGURE 3.21 Differences in the capacity of the sperm store in different species, showing major inconsistency, but related to sperm need (i.e. frequency of ejaculation over a relatively short period or to semen volume) (see text).

developed sperm store in the tail of the epididymis, but it displays a very distinct suburethral diverticulum (a fairly large pocket or pouch under the urethra), which could act as a compensatory sperm store (see Figure 4.11b, c, later).

Although, overall, the epididymis of dogs is relatively big, the tail of the duct is not well developed relative to the rest of it. This would not be appropriate, therefore, for frequent copulators because, although only a small fraction of the sperm store is emptied at each ejaculation, they will always need good back up from a large sperm store. The same applies to cats and, for this reason, frequent mountings of lionesses by male lions may not result in ejaculation, and even if orgasm is reached, the male could be firing blanks (i.e. no sperms in the ejaculated seminal plasma) with some of his multiple mountings.

By comparison, animals with a large sperm store never run out of sperms in their ejaculates, and sperms can remain viable in the sperm store even after the testes have shut down towards the end of a breeding season. Presumably, this serves as insurance in case some lonely female that is still in oestrus should cross the male's path at this time of year. This prolonged storage is seen also in deer, and viable sperms can be seen in the sperm store of bats in the middle of hibernation, long after sperm production has ceased. This supports the contention that the head and body of the epididymis have a different sort of physiology from the tail of the epididymis, the first two being directed primarily towards sperm maturation and transport and the other for prolonged sperm storage.

So, the size of the sperm store also reflects differing sexual strategies and illustrates well how the characteristics of reproductive structures in different species have evolved correspondingly.

Up to this point, discussion has been restricted to those mammals that have a scrotum. It is, therefore, appropriate now to look at those species that have no such privilege, if indeed a privilege it be.

FURTHER READING

Adams, C.E. (1962) Artificial insemination in rodents. In: *The Semen of Animals and Artificial Insemination*. Ed.: J.P. Maule. Commonwealth Agricultural Bureau, Farnham, UK.

Arpanahi, A., Brinkworth, M., Iles, D., Krawitz, A., Paranuska, A., Platts, A.E., Saiga, M., Steger, K., Tedder, P. & Miller, D. (2009) Endonuclease-sensitive regions of human spermatozoal chromatin are highly enriched in promoter CTCF binding sequences. *Genome Research*. **19**, 1338–1349.

Austin, C.R. (1965) *Fertilization*. Prentice Hall International, Inc., London.

Baumgarten, H.G., Holstein, A.F. & Rosengren, E. (1971) Arrangement, ultrastructure and adrenergic innervation of the smooth musculature of the ductuli efferentes, ductus epididymidis and ductus deferens of man. *Zeitschrift fur Zellforschung und mikroschkopische Anatomie*. **120**, 37–79.

Bedford, J.M. & Hoskins, D.D. (1990) The mammalian spermatozoon: morphology, biochemistry and physiology. In: *Marshall's Physiology of Reproduction*. Vol. 2. Ed.: G.E. Lamming. Churchill Livingstone, London and New York (this volume should be really useful, because there are several other interesting reviews in it).

Cooper, T.G. (1986) *The Epididymis, Sperm Maturation and Fertilization*. Springer Verlag, Berlin.

Cummins, J.M. & Woodall, P.F. (1985) On mammalian sperm dimensions. *Journal of Reproduction and Fertility*. **75** (1), 153–175.

de Kretser, D.M. (1990) Germ cell–Sertoli cell interaction. *Reproduction, Fertility and Development*. **2**, 225–235.

de Kretser, D.M. & Kerr, J.B. (1994) The cytology of the testis. In: *The Physiology of Reproduction*. 2nd edition. Eds: E. Knobil & J.D. Neill. Raven Press, New York.

Dott, H.M. & Utsi, M.N.P. (1971) The collection and examination of reindeer semen. *Journal of Zoology*. **164**, 419–424.

Dott, H.M. & Glover, T.D. (1999) Sperm production and delivery in mammals including man. In: *Male Fertility and Infertility*. Eds: T.D. Glover & C.L.R. Barratt. Cambridge University Press, Cambridge.

Dym, M. & Fawcett, D.W. (1970) The blood–testis barrier in the rat and the compartmentation of the seminiferous epithelium. *Biology of Reproduction*. **3**, 308–326.

Fawcett, D.W. (1966) The cell. In: *Atlas of Fine Structure*. W.B. Saunders, Philadelphia.

Fawcett, D.W. (1973) Observations on the organization of the interstitial tissue of the testis and on the occluding cell junctions in the seminiferous tubule. *Advances in Biosciences*. **10**, 83–99.

Glover, T.D. (1988) Semen analysis. In: *Advances in Clinical Andrology*. Eds: C.L.T. Barratt & I.D. Cooke. MTP Press, Lancaster, Boston, and The Hague, Dordrecht.

Hamilton, D.W. (1977) The epididymis. In: *Frontiers in Reproduction and Fertility Control*. Eds: R.O. Greep & M.A. Koblinski. MIT Press, Cambridge, MA.

Hamilton, D.W. (1990) Anatomy of mammalian male accessory reproductive organs. In: *Marshall's Physiology of Reproduction*. Vol. 2. Ed.: G.E. Lamming. Churchill Livingstone, London and New York.

Johnson, M.H. (2007) *Essential Reproduction*. 6th edition. Blackwell Publications, Malden, MA.

Jones, R. (2004) Sperm survival versus degradation in the mammalian epididymis. *Biology of Reproduction*. **71**, 1405–1411.

Jost, A. (1972) A new look at the mechanisms of sex differentiation in mammals. *Johns Hopkins Medical Journal*. **130**, 30–53.

Kormano, M. (1970) Effects of serotonin and angiotensin on testicular blood vessels in the rat. *Angiologica*. **7**, 291–295.

Kormono, M. & Penttila, A. (1968) Distribution of endogenous and administered 5-hydroxytryptamine in the rat testis and epididymis. *Annals of Medical Experimental Biology, Finland*. **46**, 468–473.

Leblond, C.P. & Clermont, Y. (1952) Definition of the stages of the cycle of the seminiferous epithelium of the rat. *Annals of the New York Academy of Sciences*. **55**, 548–573.

Mann, T. & Lutvak-Mann, C. (1981) *Male Reproductive Function and Semen.* Springer Verlag, Berlin, Heidelberg and New York.

Miller, D. & Oslermeyer, G.C. (2006) Towards a better understanding of RNA carriage by ejaculate spermatozoa. *Human Reproduction Update.* **12** (6), 757–767.

Orgebin-Crist, M.C. (1967) Sperm maturation in the rabbit epididymis. *Nature.* **216**, 816–818.

Roosen-Runge, E.C. (1977) *The Process of Spermatogenesis in Animals.* Cambridge University Press, Cambridge.

Russell, L.D., Ettlin, R.A., Sinhha Hikim, A.P. & Clegg, E.D. (1990) *Histological and Histopathological Evaluation of the Testis.* Cache River Press, Cleveland (the staging of spermatogenesis is quite complicated and really needs considerable experience, but for those who wish to pursue it, this reference is the most comprehensive and valuable one that I know).

Short, R.V. (1977) Sexual selection and the descent of man. In: *Reproduction and Evolution.* Eds: J.M. Calaby & C.H. Tindale-Biscoe (Proceedings of the Fourth Symposium on Comparative Reproduction held in Canberra, 1976), 3–19.

Susuki-Toyota, F., Ito, C., Maekawa, M., Toyama, Y. & Toshimori, K. (2010) Adhesion between plasma membrane and mitochondria with linking filaments in relation to migration of cytoplasmic droplet during epididymal maturation in guinea pig spermatozoa. *Cell Tissue Research.* **341**, 429–440.

Tesh, J.M. & Glover, T.D. (1969) Ageing of rabbit spermatozoa in the male tract and its effect on fertility. *Journal of Reproduction and Fertility.* **20**, 287–297.

Watson, P.F. (1990) Artifical insemination and the preservation of semen. In: *Marshall's Physiology of Reproduction.* Vol **2**. Ed.: G.E. Lamming. Churchill Livingstone, London and New York.

WHO Manual for the Determination of Human Semen and Sperm Cervical Interaction. (1999) 4th edition. Cambridge University Press, Cambridge.

4   The scrotum

No one can gainsay the ingenuity of some undergraduate students. When I was training as a vet, I remember medical students putting on a play and one of the cast was listed in the programme as 'Scrotum – a wrinkled old retainer'. The writers obviously recognized a feature of the scrotum, the significance of which seems to have escaped wider attention. That is, that the scrotum is indeed wrinkled.

It is popularly held that testicles descended into a scrotum because they needed to be cooled, but this belief needs much more careful examination. In the first place, the question as to why testicles descended outside the confines of the abdomen is not a legitimate one to ask, because it is shamefully Lamarkian (teleological) and not, therefore, in keeping with Darwinian principle. We can only ask how, not why. Since in the majority of, though not in all, mammals, the testicles are scrotal, the question that needs to be put is: what advantage did the evolutionary descent of testicles confer on the animals in which it happened?

In trying to answer this question, the architecture of the scrotum and its contents might provide useful information. First, the scrotum is more than a piece of stretched skin. It is modified to become exceptionally thin and has a sub-layer of elastic fibrous tissue which contains slivers of involuntary muscle (the *tunica dartos* containing *dartos muscle*). When the dartos muscle contracts, the scrotum wrinkles; when it relaxes, the scrotum is able to stretch so the wrinkles become ironed out, either partially or completely. These changes in the configuration of the scrotum are always accompanied by movements of the contained testicles. Scrotal contents are illustrated in Figure 4.1.

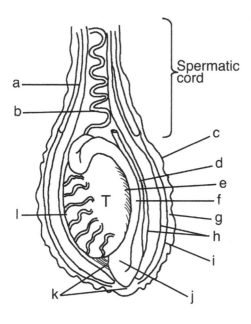

FIGURE 4.1 Contents of a scrotum. Labels: a = cremaster muscle and fascia (see text); b = helical testicular artery within the pampiniform plexus; c = wrinkled scrotal skin; d = ductus (vas) deferens; e = epididymal cleft or recess; f = epididymis; g = tunica dartos with dartos muscle (immediately under the scrotal skin); h = outer and inner layers of tunica vaginalis (peritoneum); i = spermatic and dartoic fascia (connective tissue); j = sperm store (tail of the epididymis); k = ligament of the testis and epididymis (remnant of the gubernaculum testis); l = tunic of the testis (albuginea) holding in the seminiferous tubules and the rest of the testis parenchyma.

When testicles are drawn nearer to the warmth of the abdominal region, as they are in cold conditions or in fright, the scrotum correspondingly wrinkles. When this happens due to cold environmental conditions, the surface area of the scrotum is reduced and less heat is thereby lost through radiation. But when the surrounding temperature is high, the scrotum stretches and thins, so that as much heat as possible is lost through radiation from the surface. The weight of the testicles could largely bring about the stretching, because when it is hot, they move as far as possible away from the heat of the abdomen and descend maximally. But if the scrotum's sole function were to be as a cooling mechanism for the testicle, it would never need to wrinkle. It could be argued that the scrotum is basically wrinkled in order that it can stretch for cooling, but if it only cools, why has it not evolved into a permanently stretched condition?

The answer could be that it needs this versatility in case, say, in conflict conditions, the descended testicles become overly vulnerable and need to be withdrawn nearer to the abdomen, out of harm's way.

FIGURE 4.2 Rams with their scrota artificially insulated with scrotal packs lined with Kapok.

This would mean that the scrotum serves two functions: to cool the testes and to protect them. This leaves unanswered the fact that the scrotum wrinkles and the testes ascend when it is cold. All in all, therefore, the scrotum shows itself more as a thermoregulator of the testicles than a cooler. Its protective function in aiding the withdrawal of testicles nearer to the abdomen when they are in danger looks to be an addition to the main thermoregulatory function.

So, do scrotal testicles need to be kept at a lower temperature than that of the abdomen from which they originally descended? All the evidence suggests that they do. If the scrotum of rams or bulls, for example, is insulated so that the testicles cannot lose much heat and their temperature is thereby increased (Figure 4.2), the testes malfunction and sperms in the tail of the epididymis later degenerate and become decapitated. These abnormal sperms appear in the ejaculate after about a week of treatment. This also happens when the testicles of laboratory animals are experimentally secured in the abdomen, where they are subjected to abdominal rather than scrotal temperatures (experimental cryptorchidism). However, not all the testicular cells appear to suffer at first – rather, it is only those in the later phases of spermatogenesis.

Testicular sperms and spermatids seem to be the most sensitive to increased temperature, followed by secondary spermatocytes. If the abnormal conditions are allowed to persist, primary spermatocytes are affected (they often form so-called 'phytoid cells' in which the cytoplasm seems to drop out in fixed specimens), but spermatogonia (with a few exceptions towards the end of the cycle) seem remarkably resistant, as are – on existing evidence – the other cells of the testis. It is noteworthy, therefore, that it is mainly the cells in the process of meiosis and post-meiotic cells that are particularly averse to abdominal temperature. In other words, haploid cells seem more sensitive than diploid cells.

The descent of testicles outside the abdominal environment is unique among the organs of the body, and it has been accompanied by other special features. It has taken with it part of the tunic lining the abdomen (the peritoneum), as well as pulling a flange of abdominal muscle with it (the *cremaster muscle*). This is voluntary muscle and when it contracts, it draws the testicle closer to the abdomen, or might even pull it into the abdomen in some species. It is the cremaster muscle that determines the position of the testicle, allowing it to descend when the environment is hot and ascend when it is cold.

In scrotal mammals, the embryonic testis starts life high in the upper abdominal wall (outside the peritoneum) close to the kidney, and is attached to the diaphragm. Soon it starts to migrate to the rear end of the embryonic abdomen (caudal migration). This means that its diaphragmatic attachment must give way to allow it to do so. But it is also attached to the abdominal wall by a ligament attached to the back of the abdominal wall.

This ligament pulls the testis caudally – that is, backwards towards the back of the abdomen (it is called the *gubernaculum testis*). It also prevents the testis in the embryo from ever moving forward, because it does not grow in length, so as the embryo grows, there is differential growth between the overall body of the embryo and the length of the ligament. This also helps in caudal migration. It should be pointed out that these ligaments are only jelly and are, therefore, fairly versatile in what they are able to do.

When the testis reaches the back end of the abdomen, there is a weak point on either side of the abdominal musculature which gives way and creates a slit, through which the testes drop. So there are two distinct and separate processes that occur in the embryo: one is that of caudal migration within the abdomen; the other is true testicular descent into a scrotum. The slits, which form an entry to or exit from the abdomen, are called the *inguinal canals*. In rams, and particularly in man, these canals close after the testes have passed through them, but in rabbits and hares (including the African jumping hare, remembering that this is actually a rodent rather than a hare) the canals remain wide open, so that withdrawal of the testes into the canals is easy. It is in these animals that the testes may be taken totally into the abdomen. This is not possible in rams or humans.

The passage of testes into a scrotum is mostly accomplished by the time a mammal is born, although it can be delayed. Once the testicles are *in situ*, the gubernaculum remains only as a small ligament at the lower poles of the testis and epididymis (shown in Figure 4.1).

As a result of these events, the testes in ruminants are left dangling in the scrotum, but are surrounded by peritoneum. The way this comes about is such that the testis is surrounded by two layers of the peritoneum, which together form the vaginal tunic (*tunica vaginalis*). This bilayered tunic also surrounds each ductus deferens, and immediately outside it lies the cremaster muscle. Outside this, there follows a layer of connective tissue (cremasteric fascia), surrounded by the fibrous tunica dartos, containing its dartos muscle for wrinkling the scrotum. Between the dartos and the cremasteric fascia is another layer of connective tissue, sometimes differentiated into an entity and referred to as the dartos fascia.

These details are also shown in Figure 4.1. The area of the vaginal tunic which suspends the testes between the inguinal canals and the testes themselves is known as the *spermatic cord* and it not only contains the ductus deferens, but testicular blood vessels and nerves also. In man, however, though not in the ram, the closure of the inguinal canals is such that the tunica vaginalis becomes completely separated from the abdominal peritoneum and thus becomes a separate sac in the scrotum. In the ram, the closure is not quite complete, but the passage into the abdomen is narrow and the testes are large.

THE PAMPINIFORM PLEXUS

In addition to all these features, the manner in which arterial blood is delivered to scrotal testes is unusual. The testicular artery coils in varying degrees within the spermatic cord to form a helix before entering the substance (parenchyma) of the testis. Surrounding this helix is a network of veins which lead blood away from the testis. Because the temperature of the testis is lower than that of the abdomen, it seems that this network of veins (which contains blood flowing from the testis into the abdomen) can cool the incoming blood from the abdomen, which flows in the testicular artery. It is a counter-current mechanism for heat exchange in this area, between cooler blood in the veins and warmer incoming blood in the artery. This conglomeration of blood vessels is known as the 'pampiniform plexus'. It is shown in Figure 4.3.

When the artery leaves the plexus, therefore, it will be carrying cooler blood than it originally had. It then runs on the surface of the testis, thereby allowing further heat loss from its contents through radiation. These anatomical features point to the possibility that scrotal testes have adapted to an environment in which they operate at a lower temperature than other organs of the body, and that this lower temperature is needed for meiosis to occur. However, this still leaves open the question as to why the scrotum wrinkles and testes are drawn nearer to the inguinal canal in cold conditions. On the other hand, it also begs the question of how spermatogenesis, including meiosis, can proceed in the

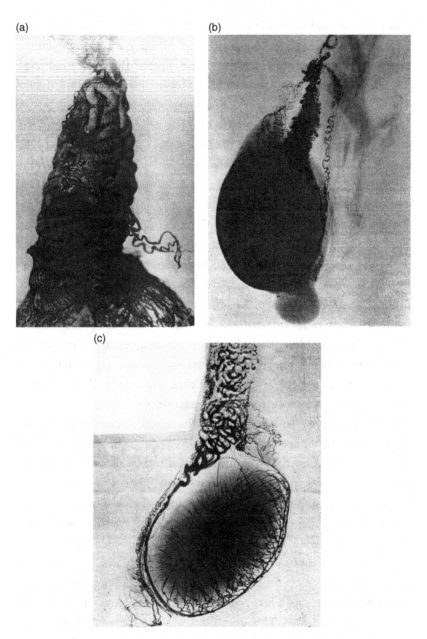

FIGURE 4.3 Helical pattern of the testicular artery in the region of the pampiniform plexus. (a) Cast showing the artery and veins in the pampiniform plexus of a bull. (b) Arteriogram of the testicular artery of a ram. Note the epididymal arteries coming off above the plexus (see text). (c) Testicular artery of a boar.

testes of birds and most testicond mammals (species whose testes are permanently retained in the abdomen), when they are constantly exposed to abdominal temperature. What is more, the abdominal temperature of birds is higher than that of mammals and testicond mammals do not have a lower than average body temperature.

If we take male vertebrates as a whole, abdominally situated testes are the norm and scrotal testes are definitely a minority condition. From this point of view, therefore, it is the testes of scrotal mammals that are out of step and it is only meiosis in their cells, it seems, that is intolerant of abdominal temperatures. There must, on this basis, be something special about meiosis in scrotal testes, and it looks as if their sensitivity to heat might be a simple adaptation to the lower temperature occasioned by their lying outside the abdomen. It has been suggested that the lower temperature of a scrotal environment might reduce the number of adverse mutations and so ensure that fewer sperms with a deficient genome are produced and released. This implies, though, that birds and testicond mammals have more mutant sperms than scrotal mammals, for which there is no evidence. Testicular descent is thus something of an enigma. But alternatives to cooling are worth examining.

For one thing, sweat glands in the scrotum are not typical or even present in all species, and it seems peculiar, if the scrotum is entirely for cooling the testes, that the scrotum of some breeds of rams are covered with a heavy layer of wool. This can be so distinct, for example in Merino rams, that farmers feel the need to shave them in very hot conditions. We should note however, that these very woolly scrota could be an unfortunate consequence of selective breeding. Even man has rather specialized hair on his scrotum (rather coarser than the rest of his body hair). Then there is a mole rat in South Africa that has a large pad of fat in the scrotum, immediately surrounding the testes. This does not seem conducive to heat loss.

Moreover, the helical nature of the incoming artery also occurs in the ovarian artery of sheep. The ovaries of ewes are abdominal and, therefore, there can be no demand for cooling – in any case, their abdominal position would mean that a counter-current heat exchange could not work. Moreover, the testes of polar bears have an exceedingly coiled testicular artery, yet a testicular cooling mechanism would hardly seem necessary in polar conditions, even if body temperature keeps other organs warm. It might be, therefore, that this curious vascular architecture primarily serves some purpose other than, or in addition to, cooling the testes. Supporting this proposition is the fact that it also gives rise to a very slow and non-pulsatile flow of blood through the testes. In some species, most notably among primates, the artery passes two or three times around the testis before it enters the parenchyma. This means that it must be exceptionally long and thus ensure an extremely sluggish blood flow. In other species the artery splits up into several branches, which could have a similar effect on specific regions of the testis.

When the temperature of the testes is elevated, one would expect a distinctive increase in blood flow, but it doesn't happen. The increase is minimal. In sharp contrast, the epididymis shows a definite increase in blood flow under such conditions, and it is interesting that the descending artery to the epididymis, as well as that going to the head of the epididymis, branches off the main testicular vessel at a point above or very close to the top of the pampiniform plexus (Figure 4.4). This means that epididymal blood flow does not effectively come under the control of the plexus. Also, if the testis is heated, it does not blush, as one might assume, but turns blue (it becomes cyanosed), which indicates that it is being starved of oxygen and that the blood flow is inadequate. Increasing the temperature of testes can only increase their metabolic activity, so they would be expected to need more oxygen than usual, but it is evident that they are unable to have it, because the pampiniform plexus prevents it.

Thus, the plexus appears to have a major controlling and restricting influence on blood flow to the scrotal testis, which might be important for its normal function. The cooling effect of the plexus might, therefore, be a secondary, if accidentally useful feature. A large lump of fat surrounding the pampiniform plexus of rabbits also seems

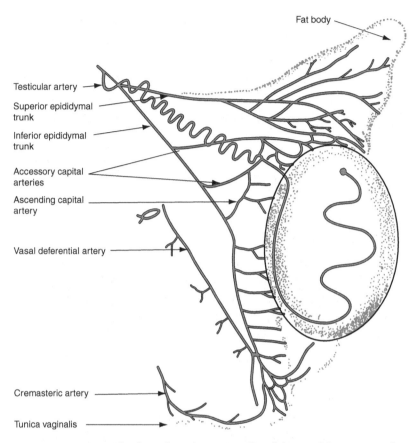

Fat body

Testicular artery

Superior epididymal
trunk

Inferior epididymal
trunk

Accessory capital
arteries

Ascending capital
artery

Vasal deferential artery

Cremasteric artery

Tunica vaginalis

FIGURE 4.4 Blood supply to the testis and epididymis of the rat. Note that
the epididymal arteries branch off the testicular artery at a point when, or
immediately after, it starts to coil – that is, above or close to the top of the
pampiniform plexus (see text). From McMillan (1954).

strange, if cooling were its main function. But obviously, the rather
odd effects of increased temperature cannot occur in cold conditions
and perhaps this explains why testes generally appear to be more
sensitive to increased rather than decreased temperature, because in
the cold, circulation will be reduced, whilst in heat, there will only be
a restrained and inadequate increase. Extreme cold will, of course,
cause damage anyway, not least due to the effects of ischaemia
(severely impaired blood flow).

Since blood flow is slowed by the helical structure of the testicular artery within the plexus, it might help the exchange of substances as well as temperature. The possibilities of oxygen and hormonal exchange have each been examined, but the results seem rather equivocal. Perhaps such measurements might usefully be made on a wider number of species. Dr Ron Hunter and I have also discussed the possibility that a coiled testicular artery might be useful in accommodating extreme testicular descent, as might be seen, for example, in bulls, by acting as a sort of spring that can be coiled or extended.

However, the scrotum, the cremaster muscle and the vascularity of the testicle all reveal a rather elaborate system of testicular thermoregulation. This points to the possibility that scrotal testicles need to be maintained within a range of temperature, rather than necessarily being held at a relatively low one. Although they seem to be more averse to increased temperature than decreased temperature, stored sperms become decapitated when the testicles are subjected to very low temperature (e.g. 0°C). Should it be, therefore, that mammalian testicles do indeed only operate satisfactorily within a range of temperature, the centre of the range will be irrelevant in different species. In other words, testes should be able to function satisfactorily at any surrounding temperature, provided it does not stray outside the range to which they have become adapted. If this were true, testicles could operate with equal facility within the abdomen or outside it, according to their evolutionary adaptation.

Whatever the ultimate explanation is, the testicles of some mammals work perfectly well inside the abdomen, whilst those of others (scrotal mammals) are unable to do so. Obviously, caudal migration of testicles within the abdomen can have little to do with temperature, and in some species other organs such as ovaries also migrate caudally. The caudal migration of the testicles in some testicond mammals, therefore, might be anticipated and it does, in fact, occur to a variable extent.

TESTICOND MAMMALS AND LOCATION
OF THEIR TESTICLES

As shown in Table 4.1, there are a surprising number of mammals with abdominal testes, but their topographical position within the abdomen is variable according to species. The testes of the elephant and the hyrax (two species which, from an evolutionary point of view, are closely related) have testes lying high in the abdomen, close to the kidney. This is more or less where they would start life in the embryo. This applies also to the testes of the aardvark, anteaters and pangolins. However, in the dugong (also closely related to the elephant and the hyrax), the testes have migrated to lie between half-way and two-thirds of the way back in the abdominal cavity. The testes of the sloth (from the same natural order as the anteaters but completely unrelated to the other three species mentioned) have also migrated similarly. Those of whales have migrated even further, so they appear to be fast approaching the inguinal canals, and those of the armadillo (again from the same natural order as anteaters and sloths) have migrated to lie just at the entrance of the inguinal canals. A helical testicular artery and a pampiniform plexus are notably absent in all these species, although the artery in armadillos undulates before entering the organ. The testes of hedgehogs have migrated higher up in the abdomen to lie immediately at the entrance to the pelvic cavity.

The testes of marine mammals other than whales are interesting. The testes of the great elephant seals and other true seal species have just popped through the inguinal canals, but are covered with a thick layer of blubber and have no scrotum. Thus, they simply form a slight bulge on the surface of the body. This is similar in the fur seals and sealions, although their bulge is a little more prominent and a distinct scrotum is evident. These animals do have a pampiniform plexus, so it looks as if the pampiniform plexus is associated with testicular descent, either partial or complete. The topographical position of testes in different mammals is illustrated in Figures 4.5–4.6. Figure 4.7 shows differences in the course of the testicular artery in

Table 4.1 *Eutherian mammals with abdominal testes ('mammalian testiconda')*

Cetaceans	Edendates (*Xenarthra*)	Hyracoids	Insectivores	Pholidotes	Proboscideans	Syrenians
All whales, dolphins and porpoises	Anteaters (giant and dwarf) Sloths (two- and three-toed) Nine-banded armadillos (and presumably all armadillos)	Rock hyraxes Tree hyraxes	Aardvarks African shrews (e.g. elephant shrews, four-toed East African shrews, yellow-backed elephant shrews) Golden moles Hedgehogs Selenodons Tenrecs	Pangolins	Elephants	Dugongs Manatees

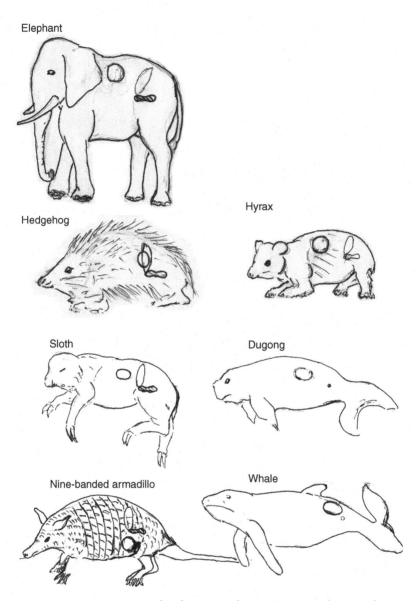

FIGURE 4.5 Topographical position of testes in testicond mammals.

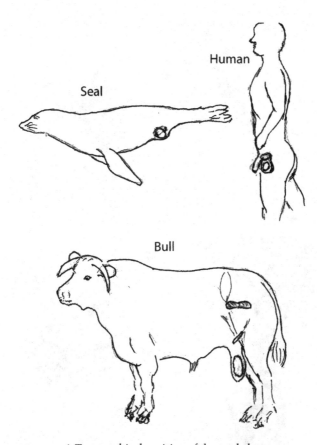

FIGURE 4.6 Topographical position of descended testes.

different species, both scrotal and testicond, emphasizing the broad association between testicular descent and the testicular artery being helical.

It is difficult to understand how the testes of elephant seals are cooled, being covered by blubber, but apparently they are, for they have been recorded as being up to 6°C below the temperature of the abdomen. Even more puzzling is that dolphins, with their abdominal testes, have a blood vessel running from their dorsal fin, which branches quite profusely and takes cooler blood to the testis through a straight testicular artery. Another straight artery runs

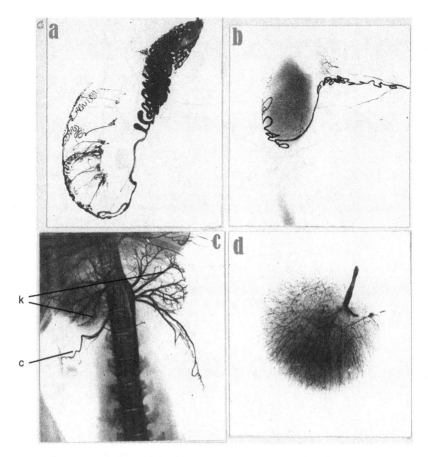

FIGURE 4.7 Arteriographs showing the arterial supply to the testes in different wild animals. (a) Giraffe (scrotal). (b) African jumping hare or springhaas (testes retractable, but scrotal). (c) Hyrax (abdominal) k=kidney; c=testicular artery. (d) Elephant (abdominal).

from the flukes (tail fins) to the testis. This provides the testis with an even lower temperature (13 °C below body temperature, apparently). This applies also to some long-finned whales, yet other whales, such as the toothed pilot whale (with a diminutive dorsal fin), have only a simple straight testicular artery, which typifies testicond mammals.

The dolphin pattern would suggest that in this species, spermatogenesis does indeed require a lower temperature than surrounding organs, but more detailed work of this kind is needed on the larger whales and a wider spectrum of whales generally, including large toothless whales and toothed killer whales. Yet the testes of other testicond mammals such as the hyrax function at body temperature and, like the elephant and dugong, have a straight testicular artery and no pampiniform plexus (see Figure 4.7 for examples).

It might be suggested that measurements of abdominal and testicular temperatures in different species would provide answers. This has been done in various species, but accurate measurements are not physically easy to make in all species, and because there are so many factors affecting body temperature, it is doubtful if one-off measurements can be considered totally reliable.

It is tempting to speculate that spermatogenic proteins in different species might differ in their susceptibility to temperature change and that this could be purely a matter of evolutionary development associated with past ecological conditions. It would be interesting to examine in more depth the nature of some of the intracellular testicular proteins in the testes of dolphins and elephant seals and compare them with those of whales. If there were to be differences, it would indicate that some testes have a need to be cool, whilst others do not. This runs slightly counter to my earlier suggestion that the range of temperature is the key, but it is quite possible that ranges could differ even among testicond species.

The testicles of man are rather interesting in that they can be considered as only just having descended into a scrotum. They are fairly close to the inguinal canals, with an abdomino-testicular temperature difference of only about 2.8 °C (in contrast to that of a bull or ram, which is about 6–8 °C) (see also Figure 4.6). Moreover, human testes are much more resistant to increased temperature than those of a ram, bull or small laboratory animal, and their epididymal sperms do not seem to mind abdominal temperature at all. For example, sperms in some men may remain viable in the ampulla (obviously at abdominal temperature) for several months, say, following vasectomy (this couldn't happen in a ram or rabbit). It is also interesting that the human testicular artery is no more coiled than that of an armadillo.

SUSCEPTIBILITY TO TEMPERATURE
OF EPIDIDYMAL SPERMS

Since there appear to be species differences in the sensitivity to temperature of spermatogenic tissue and early (immature) sperms, it is reasonable to ask if this applies to mature sperms in the sperm stores. Since the epididymal vasculature does not seem to be under the control of the pampiniform plexus, it might be expected that the epididymis of all scrotal mammals might function normally when its temperature is elevated and that stored mature sperms are unresponsive. However, this is not the case in commonly used laboratory animals. Like haploid cells within the testes, immature sperms in the epididymis of laboratory animals respond adversely to increased temperature within a matter of hours, whilst mature ones stored in the tail of the epididymis do not respond for several days. Clearly, mature sperms are more resistant to heat than immature ones. Nevertheless, except in man, they ultimately succumb.

All this indicates that in most scrotal mammals, blood flow apart, there must be some intrinsic effect of increased temperature on post-meiotic cells, including the sperms themselves, or on the testicular and epididymal milieu. It has been suggested that the heat susceptibility of epididymal sperms points to their being the main cause of testicular descent, so that the epididymis would be a 'prime mover' in the evolutionary process of descent. However, the relative resistance to increased temperature of sperms stored in the tail of the epididymis of most scrotal mammals (and throughout the epididymis of man) rather invalidates this theory, which in any case is another teleological one. More than this, stored sperms in testicond mammals are also quite content with abdominal or pelvic temperature.

After ejaculation, too, sperms seem to become relatively indifferent to small increases in temperature, be it in a container or in the female tract. They normally become decapitated and die off fairly quickly in the vagina, but higher up the tract they can remain viable for quite long periods in some species, and we have already seen that they can pass into the peritoneal cavity and still be alive and kicking. The fragility of sperms in the male reproductive tract of scrotal mammals in response to increased temperature must, therefore, be very much associated with their immediate environment. It could also be that seminal plasma might initially afford some protection against slight temperature change. If this were so, then it cannot be total, because in the other direction, ejaculated sperms in some species can be 'shocked' if subjected to very low temperature. It is interesting, though, that before ejaculation sperms are immotile, and in conditions of low oxygen tension, whilst precisely opposite conditions prevail after ejaculation.

It must be that in most scrotal mammals, with the possible exception of man, survival factors within the testis and epididymis are thermolabile, whilst in testicond mammals they are either chemically different or absent.

Each sperm store in the hyrax is remotely situated from the abdominal testes and lies as a discrete region of the epididymis within the pelvic cavity, as shown in Figure 4.8 (this is also characteristic of the store in African shrews). Whether or not this region is an homology of the ampulla of the ductus deferens or the tail of the epididymis (which it looks like) is uncertain, but even if its position affords a slightly lower temperature than that of the abdomen, it surely offers a more important mechanical advantage in ejaculation by having sperms stored close to the point of exit – that is, near to the pelvic urethra. Were it not so, and stored sperms lay near to the testis, as they do in scrotal mammals, they would have to make an unacceptably long journey before being able to exit the male tract when needed (Figure 4.8).

Although sperm storage in the elephant is more diffuse than in the hyrax, there is a slight enlargement of the excurrent duct towards its terminal end as if there is some extra sperm storage at that point (Figure 4.9). In testicond species, whose testes have migrated

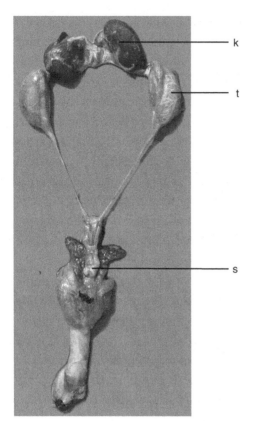

FIGURE 4.8 Reproductive tract of a rock hyrax. Note the remotely situated sperm store. Labels: k=kidney; t=testis; s=sperm store.

caudally, however, the epididymis has become more coiled even than that of the elephant, and as in scrotal mammals, the sperm store has come to lie adjacent to, if extending partly beyond (as in sloths) the testis. However, because of migration of the testes, the sperm store in these species lies closer to the exterior anyway, so that mature sperms will have a shorter path to exit than would those of the hyraxes and African shrews if their sperm store were close to the testis. The end of the tube beyond the sperm store is very short in the hyrax, shrews and elephant; it is rather longer in testicond species where the testis has migrated, and longer still in scrotal mammals (ductus deferens). But when a scrotal mammal ejaculates, the testes are drawn closer to the inguinal canals, and since the sperm store is adjacent to the testis, the passageway into the pelvic urethra and

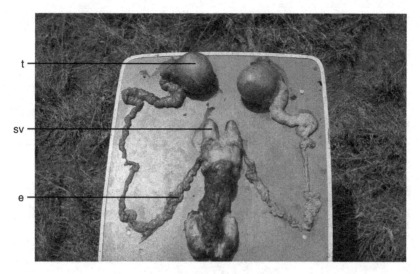

FIGURE 4.9 Reproductive tract of an elephant. Clearly the sperm store is more diffuse than in the hyrax. Labels: t=testis; sv=seminal vesicles; e=epididymis.

thence to the exterior is thereby shortened and the vasa deferentia must shorten too, presumably by a degree of contraction.

It is interesting to note, though, that the degree of musculature surrounding the epithelium of the ductus deferens in different species appears to correspond to the distance sperms have to travel from the sperm store to the exterior during ejaculation. Thus, the terminal part of the tract in hyraxes and elephants has little muscle (not far to go), testicond mammals with partial migration of the testicles (e.g. dugong and sloth) have a little more, whilst scrotal mammals have an exceptionally substantial muscular layer, providing a forceful pump for the delivery of sperms into the pelvic urethra during ejaculation. This is shown in Figure 4.10.

It has obviously been important, therefore, for sperm stores to have the easiest access possible to the exterior. This could have been a benefit to some species, resulting from the evolutionary development of caudal migration of testicles within the abdominal cavity. It is much more plausible than any notion that abdominal migration reflects a frantic effort on the part of testicles to get out of the

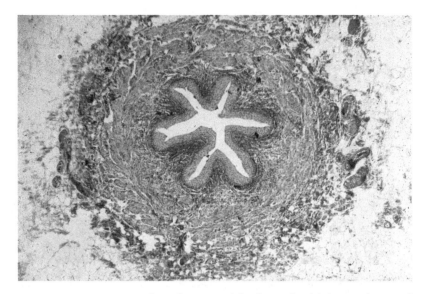

FIGURE 4.10 Histological section of the ductus (vas) deferens of a scrotal
mammal, illustrating its massive muscular layer (involuntary muscle).

abdominal environment in order to cool off. When caudal migration
was extreme, it is easy to imagine how testes might accidentally have
slipped through the inguinal canals and thereby become functionally
modified.

But the sperm store in testicond mammals is not always as
discrete as it is in the rock hyrax, and it is evident that a much greater
length of duct serves as a sperm store, for example, in the elephant.
In this regard, it is also interesting that some birds have a discrete
sperm store like the hyrax (seminal glomus), but others, such as the
domestic cock, store sperms throughout the length of the epididymis.
It is apparent that, in mammals, the storage system varies similarly,
so that the duct extending from the testis to the exterior (mesonephric duct) may mostly be straight, as in hyraxes and pangolins, but in
the giant anteater, it is even more coiled than in the elephant, and
doubtless sperms are stored here throughout most of the length of the
duct. In this respect, therefore, the anteater seems to be rather like a
domestic cock. By contrast, the epididymis of the nine-banded

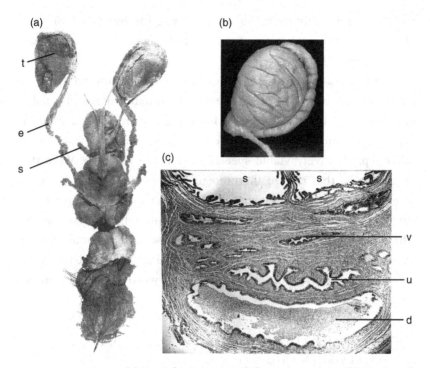

FIGURE 4.11 (a) Reproductive tract of the giant anteater. (b) Testis and epididymis of the nine-banded armadillo (note diminutive sperm store at the level of the tail of the epididymis). (c) Histological section through the sub-urethral diverticulum of the nine-banded armadillo, containing sperms. Labels: t = testis; e = epididymis; s = seminal vesicles; u = pelvic urethra; d = sub-urethral diverticulum with sperms in it; v = vas deferens.

armadillo resembles more closely that of scrotal mammals. However, the position of its testes being so close to the inguinal canals makes it look as if this animal was poised to be a scrotal mammal during the course of evolution, but never quite made it!

Each kind of epididymis is shown in Figure 4.11. The pictures shown in Figure 4.11(b) and 4.11(c) have already been referred to in Chapter 3 in relation to sperm stores. It is interesting to note that there is a vague relationship between a discrete sperm store and intermittent reproduction. Seasonally breeding testicond mammals, for example, have more discrete sperm stores than those species that have no seasons (presumably because they need a good supply close to hand for a relatively short period – compare the hyrax and the elephant). This seems to follow a pattern in birds and, to some extent, scrotal mammals.

If it is assumed that scrotal testicles have become accustomed to a cooler environment occasioned by their being outside the abdomen, the question remains as to what was gained by their pushing or being pulled into a scrotum in the first place, when the testicles of other reproductively viable species have failed to do so. Were the answer to be that careful thermoregulation conferred an advantage by ensuring that testicles operated within a narrow range of temperature, can it be assumed that testicond mammals are less fertile? There is nothing to suggest that this is the case, so how could delicate thermoregulation have been important? The possibility put forward below is, of course, purely conjectural, but is worth thinking about as an explanation.

EVOLUTION OF TESTICULAR DESCENT: A POSSIBLE SCENARIO

Almost 200 million years ago, after the dominant Jurassic reptiles became extinct, archaic mammals made their appearance on the scene (Tertiary period of the Cenozoic era), and about 20 million years later, with the arrival of the epoch known as Eocene, species akin to modern mammals became ever more prominent. Before this, one imagines that many features of early mammals would more closely resemble those of their reptilian ancestors than of contemporary mammals. Thus, it is possible that these earliest of mammals might not only have had abdominal testes, but could well have had poor internal body thermoregulation as well.

Amphibians and reptiles are known as being ectothermic (poikilothermic). This means their body temperature is governed by the temperature outside. If they lie in the sun, they warm up, and if they lie in the shade, they cool down. This is not the most efficient form of thermoregulation of the body and must result in sharp variations in body temperature. It is argued here, therefore, that early mammals (say, in the Palaeocene, long before the later epochs of Oligocene or Miocene), might also have been ectothermic, or virtually so, and that this could have resulted in testicular temperature being regularly subjected to temperatures outside their range of comfort. However, in some of the more modern mammals, the range could be maintained through accidental descent of the testicles into a distinctly thermoregulatory scrotum.

Later, new genera of mammals having much better body thermoregulation (i.e. becoming almost endothermic or homeothermic) must have appeared on the scene. This homeothermy had probably already happened in birds and might even have begun in some dinosaurs. However, it would mean that there could be mammals with a scrotum, but that yet were also endothermic (namely, having efficient internal regulation of body temperature). It does not mean to say that they started out like that. Testicular descent could have preceded efficient endothermy. If this were the case, then descent of testes into a scrotum would ultimately have conferred no advantage. But it happened, and descended testes could have already adapted to the scrotal position by the time endothermy developed in their owners. They would thus function at a temperature lower than body temperature.

Sea water (or water of any kind) conducts heat away from the body more expeditiously than air, so blubber is very necessary if an aquatic mammal is to be kept warm. It is very efficient, so whales, like birds, could have developed endothermia from the start (in this context, it is particularly worth remembering that whales are descended from land mammals). If this were true, their abdominal testes would be held within a fairly narrow temperature range anyway. Minus the blubber, this could also have happened later in the sloth, which is always a bit late for everything!

By contrast, some land testicond mammals such as the rock hyrax and the east African shrews have remained relatively ectothermic. Theoretically, if the temperature range theory is correct, they would be expected to have a problem with prolonged testicular function, because it could easily move outside their comfort range. But like dugongs, their testes do not function at all for most of the year, and a short mating season characterized by an equally short, but enormous, spurt in sperm production doubtless overcomes the problem. This does not apply to the elephant, but its leisurely breeding pattern could be accompanied by better body thermoregulation. After all, the whole body of an elephant is covered by a sort of scrotum, and taking the size of the animal into account, there is a huge surface area for heat loss anyway! Moreover, the ears of elephants are particularly efficient thermoregulators.

It has been suggested in the distant past that when whales moved from land to the ocean, they could have withdrawn their testes perhaps for streamlining (or because of sharks!), but there is no need to postulate that their ancestors had scrotal testes in the first place. Their abdominally migrated testes indicate that, if anything, the testes of whales have been on their way down rather than having been withdrawn.

Overall, the evidence indicates that most mammals attempt to keep the temperature of their testes stable, and those that can't have developed their own special devices for coping with the problem.

So is a scrotum really necessary? It is plain for all to see that, truism it may be, but in contemporary mammalian species, it is needed by those that have one and is not by those that do not. But an important possibility raised in this chapter is that testicular descent into a scrotum might be unrelated to the actual temperature of testicles (i.e. including both testes and epididymides) per se and that the stability of their temperature might be, and might in the past have been, much more significant.

A scrotum and its contents are naturally regarded as part of the external genitals of a male, so having described most of the organs of reproduction, it is appropriate now to discuss how they are used in the mating process. As far as mammals are concerned, we shall see that another external genital organ, the penis, is all-important in this.

FURTHER READING

Backhouse, K. M. (1964) The gubernaculum testis Hunteri: testicular descent and maldescent. *Annals of the Royal College of Surgeons of England*. **35**, 15–33.

Backhouse, K. M. & Butler, H. (1960) The gubernaculum testis of the pig (*Sus scrofa*). *Journal of Anatomy*. **94**, 107–120.

Bryden, M. M. (1967) Testicular temperature of the southern elephant seal *Mirounga leonine* (Linn). *Journal of Reproduction and Fertility*. **13**, 583–584.

Bryden, M. M., Marsh, H. & Shaughnessy, P. (1998) *Dugongs, Whales and Dolphins*. Allen & Unwin, St Leonards, NSW.

Carrick, F. R. & Setchell, B. P. (1977) The evolution of the scrotum. In: *Reproduction and Evolution*. Eds: J. H. Callaby & C. H. Tyndale-Biscoe. Australian Academy of Sciences, Canberra.

Chang, M. C. (1943) Disintegration of epididymal spermatozoa by the application of ice to the scrotal testis. *British Journal of Experimental Biology*. **20**, 16–23.

Crew, F. A. E. (1922) A suggestion as to the aspermatic condition of the imperfectly descended testis. *Journal of Anatomy*. **56**, 98–106.

Cross, B. A. & Silver, I. A. (1962) Some factors affecting oxygen tension in the brain and other organs. *Proceedings of the Royal Society of London. Series B*. **156**, 483–499.

Free, M. J. (1977) Blood supply to the testis and local exchange and transport of hormones in the male reproductive tract. In: *The Testis*. Eds: A. D. Johnson & W. R. Gomes. Academic Press, New York.

Glover, T. D. (1955) Some effects of scrotal insulation on the semen of rams. *Proceedings of the Society for the Study of Fertility*. **7**, 66–76.

Glover, T. D. (1960) Spermatozoa from the isolated cauda epididymidis of rabbits and some aspects of artificial cryptorchidism. *Journal of Reproduction and Fertility*. **1**, 121–129.

Glover, T. D. (1966) The influence of temperature on flow of blood in the testis and scrotum of rats. *Proceedings of the Royal Society of Medicine*. **59**, 765–766.

Glover, T. D. (1973) Aspects of sperm production in some East African mammals. *Journal of Reproduction and Fertility*. **35**, 45–53.

Glover, T. D. & Sale, J. B. (1968) The reproductive system of male rock hyrax (*Procavia* and *Heterohyrax*). *Journal of Zoology*. **156**, 351–362.

Gorman, M. (1976) Seasonal changes in the reproductive pattern of feral *Herpestes auropunctatus* (Carnivora: Viverridae) in the Fijian Islands. *Journal of Zoology. London*. **178**, 237–246.

Grassé, P. P. (1955) Ordres des hyracoides on hyraciens. In: *Traite de Zoologie*. Ed.: P. P. Grassé. Masson, Paris.

Harrison, R. G. (1949) The comparative anatomy of the blood supply to the mammalian testis. *Proceedings of the Zoological Society of London*. **19**, 325–343.

Harrison, R. G. & Weiner, J. S. (1949) Vascular patterns of the mammalian testis and their functional significance. *Journal of Experimental Biology*. **26**, 304–316.

Jones, R. C. & Djakiew, D. (1978) The role of the excurrent ducts from the testes of testicond mammals. *Australian Zoologist*. **20** (1), 201–210.

Marsh, H., Heinsohn, G. E. & Glover, T. D. (1984) Changes in the reproductive organs of the male dugong (*Sirenia dugongidae*) with age and reproductive activity. *Australian Journal of Zoology*. **32**, 721–742.

McMillan, E. W. (1954) Observations on the isolated vaso-epididymal loop and the effects of sub capital epididymal obstructions. *Studies on Fertility*. **1**, 57–64.

Romer, A. S. & Parsons, T. S. (1986) *The Vertebrate Body*. 6th edition. Saunders College Publications, Philadelphia.

Rommel, S. A., Pabat, D. A., McLellan, W. A., Mead, J. G. & Potter, C. W. (1992) Anatomical evidence of a countercurrent heat exchange associated with dolphin testes. *Anatomical Record*. **232**, 150–156.

Rommel, S. A., Pabat, D. A., McLellan, W. A., Williams, T. M. & Friedl, W. A. (1994) Temperature regulation of the testes of the bottle nosed dolphin (*Tursiops truncatus*): evidence from colonic temperatures. *Journal of Comparative Physiology B*. **164**, 130–134.

Setchell, B. P. (1978) *The Mammalian Testis*. Paul Elek, London.

Setchell, B. P. & Waites, G. M. (1964) Blood flow and the uptake of glucose and oxygen in the testis and epididymis of the ram. *Journal of Physiology*. **171**, 411–425.

Short, R. V. (1997) The testis: the witness of the mating system, the site of mutation and the engine of desire. *Acta Pediatrica: Supplement*. **322**, 3–7.

Waites, G. M. (1980) Functional relationships of the mammalian testis and epididymis. *Australian Journal of Biological Sciences*. **33**, 355–370.

Waites, G. M. & Setchell, B. P. (1964) Effect of local heating on blood flow and metabolism of the testes of the conscious ram. *Journal of Reproduction and Fertility*. **8**, 339–349.

Weber, M. (1898) *Studien über Säugetiere*. Part 2. Gustav Fischer Verlag, Jena.

5 The delivery

As a male becomes sexually excited and ready to mount his female, his sexual equipment is operable and semen is poised for discharge. But wait! There is yet another all-important dimension to be considered. Before excitement kicks in, the penis is soft, flexible and altogether floppy (flaccid) in many species. This is no good for intromission. First, it must stiffen, else it is likely to bend on impact. This hardening and enlargement of the penis is *erection*, and it is induced by the sight of a female in heat, smell, or the anticipation of coitus. As we shall see later, in man, it can occur spontaneously or through imagination or fantasy.

ERECTION

Messages run from the brain, through the spinal cord, to be carried in nerves near to the end of the cord at the level of the pelvis. These pelvic nerves (*nervi erigentes*) end in the arterial blood vessels of the penis and, when stimulated, cause them to dilate. Unlike most arteries, which end in capillaries, these end in large cavernous spaces in the shaft of the penis, which become filled with blood and endow the organ with a size and stiffness, the like of which, if for the first time, it has not previously experienced. The muscles in the walls of the blood vessels, which cause them to dilate, respond involuntarily, so erection is classed as being an involuntary or parasympathetic reflex. The blood vessels are quite long and are thus seen to be variably coiled (*helicine arteries of the penis*).

Presently, adrenaline begins to rush around the body, the heart beats faster, semen (containing both sperms and seminal plasma) is delivered into the pelvic region of the male tract by other spinal reflexes (*seminal emission*) and exuberant accessory organs produce

FIGURE 5.1 (a) Thoroughbred stallion about to cover a mare. It should not be necessary to guide a stallion's penis as some handlers are wont to do. It is not good practice and can deter some stallions. (b) Male black rhinoceros about to begin copulating. © Photolibrary.com. (c) African elephant mounting a female. © Photolibrary.com.

FIGURE 5.2 Stallion covering a mare (pelvic thrusting) (courtesy W.R. Allen)

pre-coital fluid at the end of the penis (this usually comes from the urethral glands). Nerves carrying messages for seminal emission are so-called sympathetic nerves and emerge from the spinal cord at the level of the lower thoracic and upper lumbar regions of the cord. They pass through small conglomerations of nervous tissue called *ganglia*, and the post-ganglionic fibres run to the accessory organs, including the tail of the epididymis, so that both sperms and seminal plasma enter the pelvic urethra and are ready to go. The male is now ready to mount and intromission (insertion of the penis) follows (see Figures 5.1 and 5.2 showing different species about to intromit).

PELVIC THRUSTING

Immediately after intromission, males begin pelvic thrusting. This involves contraction of the gluteal muscles (although in quadrupeds it is primarily the hamstrings) and muscles of the back, above the vertebrae (epaxial muscles), alternating with contraction of the deeper (hypaxial) back muscles (and probably the abdominal muscles), to produce a rhythmic forward thrusting or oscillation of the pelvis. This is typical of mammalian copulatory activity. The signal or stimulus for this activity, like that which brings about the ultimate discharge of semen from the end of the penis (*ejaculation*), originates from sensations in the penis itself. When a penis becomes erect, its end, or *glans*, becomes extra sensitive and is said to be in a state of *tumescence*. This is why the two activities of thrusting and ejaculation do not occur when the penis is flaccid, because in the flaccid state, the sensitivity of the nerve endings is insufficiently heightened. When they are heightened, impulses picked up from the end of the sensitized penis are carried to the spinal cord (sensory fibres) and responding impulses (carried in the fibres of motor nerves) are then directed to muscles for thrusting and, when orgasm is reached, to penile muscles for ejaculation. The stimulation of ejaculatory muscles is known as a simple spinal reflex, and because the motor nerves lead to skeletal or voluntary muscles (e.g. the bulbocavernosus or main ejaculatory muscle) it is also referred to as a somatic reflex.

Emission continues during most of the pelvic thrusting phase, and a high level of circulating adrenaline augments the process. Thus, a ram or rabbit that is highly sexually excited (more adrenaline) produces more sperms in his ejaculate (and probably a slightly larger volume of semen) than one that is indifferent about copulating or one that ejaculates so quickly that it has no time to get worked up. Second ejaculates are often of better quality than the first ones if they occur in fairly quick succession, again because circulating adrenaline has not yet died down and is likely to be at a higher level than at the first ejaculation.

As male orgasm approaches, the thrusts become more rapid and the final climax is accompanied by ejaculation. In most species (though we shall see that there are exceptions), erection, and hence tumescence, quickly disappear after ejaculation and the male dismounts. The party is now over till the next time. The nerves involved in these processes are shown in Figure 5.3.

Like erection of the penis, pelvic thrusting appears to have some input from testosterone, but neither is completely controlled

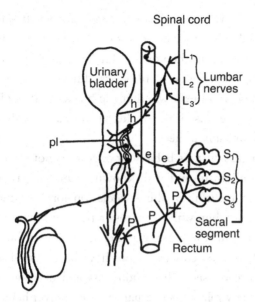

FIGURE 5.3 Nerves involved in male coital activity. Labels: h = hypogastric (presacral) nerve – seminal emission; e = pelvic nerve (nervus erigens) – erection (both these are involuntary (autonomic nerves); P = pudendal nerve extending from the dorsal nerve of the penis – ejaculation (this is a somatic nerve since its fibres pass to and from voluntary (ejaculatory) muscle; this is the one that needs either friction or specific temperature for stimulation from a tumescent penis; pl = pelvic plexus of nerves through which impulses from the hypogastric nerves and the nervi erigentes pass. Example taken from the distribution of nerves in the cat. Redrawn from Semans and Langworthy (1938).

by it. For instance, before they can have any significant levels of circulating testosterone, human male babies occasionally experience spontaneous erections. Also, if stallions at stud are castrated, they will often continue to mount mares, displaying a full erection, for years afterwards. Although this cannot be the case with babies, the observation on castrated stallions suggests that imprinting comes into the picture also.

But female sex hormone seems to play a part in male coital behaviour as well. Female animals in heat often display male behaviour. It can be something of an embarrassment at an afternoon tea party, when the hostess' bitch suddenly grasps the leg of one of the guests and starts pelvic thrusting! When cows are in oestrus they frequently mount other cows and pelvic thrust. This is used by farmers to recognize when a cow is in heat. In the north of England, the behaviour is known as 'hockering' (doubtless a word of Scandinavian origin). It is sometimes hard to know, though, whether the hockerer or the hockered is the one in heat, because if the hockered cow stands still when mounted, she too is probably in heat.

It appears that this curious behaviour of cows is not altogether species-specific. A delightful and well-known animal behaviourist once announced at a meeting of the Association for the Study of Animal Behaviour that when walking across a field on his farm, one of his cows mounted him. A wag in the audience impiously, but good humouredly, teased that the cow could not have been very discerning! But we have already seen that androgens and oestrogens are chemically quite closely related, so perhaps this crossover behaviour of the sexes is not so surprising after all.

EJACULATION

It is evident that the spinal reflex that gives rise to ejaculation originates by stimulation of the glans of an erect penis. It results in a signal being sent back along the sensory fibres of the main nerve of the penis to the spinal cord; motor fibres (instructing the appropriate ejaculatory muscles to contract) pass out of the cord in the same nerve. This nerve runs on the upper surface of the penis (the 'dorsal nerve'), but it has branches going off to the skeletal muscles of ejaculation situated at the base of the penis. In some animals, temperature is an all-important factor in releasing the ejaculatory reflex, whilst in others it is friction on the penis during coitus that is the vital stimulus.

This difference is highlighted when semen is collected with an artificial vagina. The temperature of the water within an artificial vagina is an essential factor in achieving ejaculation in buck rabbits, rams and, perhaps especially, bulls. It needs to be adjusted to about 37°C, although it varies slightly in individuals. In boars and dogs it is quite different –the temperature of an artificial vagina used with

FIGURE 5.4 Artificial vagina for collecting semen from a rabbit. It is important to fill the cylinder with water at the correct temperature, which varies to some extent with individual bucks.

these animals is unimportant in releasing the ejaculation reflex – it can be induced simply by means of manual stimulation of the penis. An artificial vagina for rabbits is illustrated in Figure 5.4.

Thus, it is apparent that the presence of a female is not an essential requirement for a male animal to ejaculate. But there are grades of sexual excitation before and during copulation which are not governed exclusively by mechanical factors. It has been shown, for example, that in rabbits, bucks are most excited when they are presented with a live female. So if rabbit semen is to be collected using an artificial vagina, the ideal method is to hold it underneath a doe. However, this is a slightly clumsy exercise, so the next best performance of the buck is obtained by the operator placing the skin of a female rabbit over his hand as he or she holds the artificial vagina. The third grade of excitation is elicited with an unadorned hand of an operator holding the artificial vagina, but the least effective inducer of copulation is an inanimate object. Hence, it looks as if, whatever the species is (even a human will do), animation is important in inducing an animal to mount.

This seems to apply less to boars – they need no encouragement to mount an inanimate object and will quite freely mount an artificial sow (see Figure 5.5). Primates, too, seem to be versatile, and semen has been collected from monkeys by means of an artificial vagina. Monkeys are frequently observed masturbating, so theoretically, this should not be a problem for them. However, I once advised a group of

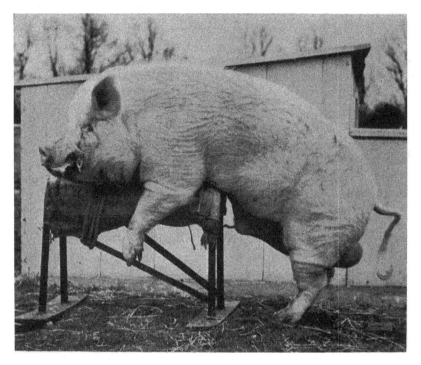

FIGURE 5.5 A boar serving a dummy sow. From Walton (1968).

zookeepers to use this method of semen collection with a monkey and it turned out that the monkey ate the artificial vagina! It can't have been very impressed.

Electro-ejaculation is a more reliable means of collecting semen, especially in wild animals. It is a pretty blunderbuss way of obtaining semen, because it sends shock waves not only onto some of the nerves in the area, but also onto the accessory organs themselves. Nevertheless, it works and methods have become much more refined since it was first used. With bulls, semen can be obtained by this means without stimulating erection, but in smaller ruminants such as the ram or deer, erection usually also occurs. We once tried the technique on a paraplegic man and he ejaculated quite unexpectedly, both to his and our surprise. So, ejaculation can occur without erection, because they are separate and different reflexes.

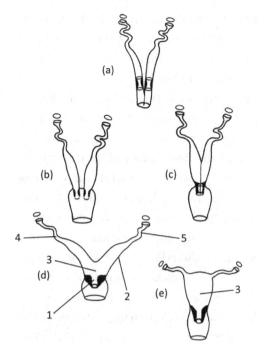

FIGURE 5.6 Comparative structure of female tracts (see text). (a) Vagina duplex. (b) Uterus duplex. (c) Uterus bipartitus (bipartite uterus). (d) Uterus bicornis (bicornuate uterus). (e) Uterus simplex. Labels: 1 = cervix; 2 = horn of the uterus (cornu); 3 = body of the uterus; 4 = utero-tubal junction; 5 = Fallopian tube.

However, in coitus, the routine of delivering semen is through the sequential processes of erection, mounting and intromission, pelvic oscillation, seminal emission and, finally, ejaculation. This sequence is common to all mammals. But where is the semen to be placed within the female tract? Each species of mammal has its own idea about this and so each has its own method of coitus. To detail these methods, it is necessary to examine the female tract of different species.

FEMALE TRACTS

Female tracts are illustrated in Figure 5.6. Most marsupials have two vaginae (*duplex vagina*), each with a uterine horn extending from it. As a result, they have not one, but two cervices and two uterine horns (*duplex uterus*). Some marsupials, such as the kangaroo, have three vaginae (two lateral vaginae that transport the sperm and a median birth canal), but in the context of this evolutionary principle, it has two cervices. Incidentally, a male marsupial has a forked penis, but the two cervices are not there to accommodate each prong. Some eutherian mammals such as some rodents also have this kind of uterus, including the two cervices. However, in the majority of eutherian mammals, the two uterine horns have converged, so that in cats and most rodents, a single cervix is present, but it is divided into two parts by a central flange of tissue or septum (*bipartite uterus*). Further convergence not only produces a single cervix, but a body of the uterus, from which the two horns emerge (*bicornuate uterus*). This is typical of domestic species and wild ruminants as well. Further convergence of the uterine horns yields a *simplex uterus*. This kind of tract typifies primates, including women. Uterine horns have disappeared here and the tract consists only of a large body of the uterus connecting directly to the Fallopian tubes. Mares' uteri have been classified as simplex too, but I think this is debatable and I personally think of them as being bicornuate.

In some species such as the horse, primates (including the human female), rabbits and rodents, the cervix protrudes into the vagina centrally, so that the blind front end of the vagina (*fornix*) is more or less equidistant all round the cervix or cervices. The cervix of sows is also centrally placed, but does not protrude into the vagina. The cervix of ruminants, however, juts out just above the floor of the vagina, so there is no ventral part to the fornix. Taking copulatory tactics into account, we shall see that this is a convenient arrangement for ruminants. Contrary to this arrangement, in dogs, there is no dorsal fornix (see Figure 5.13). The entrance to the cervix from the vagina and its exit into the uterus can be referred to by using the terms 'external os' and 'internal os' of the cervix, respectively (presumably because the cervix feels rigid, rather like a bone).

PENIS

The intromittent organ of the male is the penis or phallus. It is fashioned to suit the mode of coitus in each species.

There are two basic types of penis, classified according to the way in which they become erect. The *vascular penis*, such as that of stallions,

FIGURE 5.7 Extended penis of the elephant (not erect). I am given to understand that only the elephant can step on his penis! Courtesy of W.R. Allen.

elephants (Figure 5.7), hippos, men, dogs, cats, rabbits, rodents, bears, seals, walruses, sea cows and others, erects purely by means of the cavernous spaces in its shaft filling with blood. This stiffens the sinews, as it were – the penis becomes turgid with blood and enlarges. Characteristically, before erection, this kind of penis is flaccid and is only to be seen during urination or immediately after coitus (Figure 5.7). Normally, it is kept discreetly hidden within the prepuce (foreskin).

As a matter of interest, the prepuce of stallions has a double internal fold in it and prepucial secretion ('smegma') regularly accumulates within. Sometimes it can be heard gurgling in there, especially when a horse is trotting on a hard road. To avoid this, or in the interests of hygiene in a stallion at stud, dessicated smegma can be washed off a contaminated penis by pulling it out of its prepuce. The procedure is known as 'drawing the yard'. How wild stallions cope I am not sure, but there is always the chance that if it is not attended to it might cause inflammation ('balanitis'). This can happen also in uncircumcised men, especially if they have a very long prepuce or foreskin.

The second form of penis is called a *fibro-elastic penis*. This is possessed by all ruminants, but also by boars, rhinos (Figure 5.1) and whales. Doubtless other species also have this type of penis. This kind of penis is never flaccid and the erectile tissue (cavernous spaces) is less well developed. However, the penis is held within the prepuce by virtue of an S-shaped bend in it. When the penis erects, blood pours into what erectile tissue there is and enlarges the penis somewhat, but by making it turgid, the S-shaped bend is straightened out and the penis extends, to be thrust forward and protrude from the prepuce. Erection here, therefore, is a slightly less dramatic affair than with a vascular penis, for it is not so much an increase in size and stiffness that results, so much as protrusion. Different penes among domesticated mammals are illustrated in Figures 5.8–5.12.

PENIS AND THE MODE OF COITUS

Having produced semen, the male of each species has to decide where to put it. You might think this is obvious, but it is not as simple as it seems, because the method of coitus adopted by different species varies. These are shown in Figure 5.13 and illustrated in Table 5.1. Normally, the entrance to the vagina (the *vestibule*) points downwards from the main body of the vagina. This would make penetration difficult, but a receptive female raises her hindquarters in varying degrees, making her back concave (*lordosis*) and this facilitates easier entry of the penis.

STALLIONS

Like elephants, stallions have a large vascular penis. When erect, the glans in particular swells up because it has two separate sources of blood supply (Figure 5.8). More blood flows into it, therefore, than into the rest of the penis and it expands to about the size of a saucer. This is very useful, because the stallion pushes his penis firmly against the cervical orifice during coitus to ensure that most of the semen is deposited directly into the cervix. A little bit of the

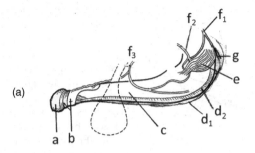

(a)

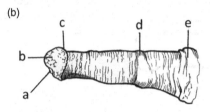

(b)

FIGURE 5.8 Penis of the stallion. (a) Non-erect penis. Labels: a = glans penis, corona glandis; b = collum glandis and dorsal process; c = shaft of the penis (contains most of the cavernous tissue); d_1 = retractor penis muscle (withdraws the penis after erection); d_2 = main muscle for ejaculation (bulbocavernosus muscle); e = muscle for positioning the penis and aiding erection (ischiocavernosus muscle); f_1 and f_2 = upper arteries supplying blood to the penis; g = artery going to the 'bulb' of the penis (the bulb is another mass of cavernous tissue that immediately surrounds the urethra and thus passes along the length of the penis and supplies extra blood to the glans) (see text). (b) Erect penis. a = urethral process jutting out of a lower indentation of the glans penis (fossa glandis); b = glans; c = corona; d = preputial ring; e = fold of penis everted during erection (the prepuce of stallions has two folds in it to sheath the penis – d and e represent these two folds).

urethra juts out in front of the glans and enters the cervix itself, just as a bit of extra insurance (Figure 5.13). There is some backflow because of the large volume of ejaculate, but some horse breeders claim that the cervix dips into the residue and sucks it up. I think this could be fanciful and perhaps a little anthropomorphic. Total backflow is prevented by the gel (discussed earlier) when it is present. Some scientists believe the gel is more concerned with preventing semen from other males entering the tract, but in horses this is likely, even with their distant ancestors, to be a secondary

(a)

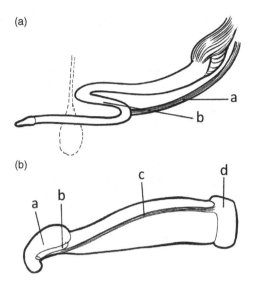

(b)

FIGURE 5.9 Penis of the bull. (a) Non-erect penis. Labels: a = retractor penis muscle (pulls the S-shaped bend of the penis – and thus the penis itself – back into the prepuce after erection); b = sigmoid flexure of penis (S-shaped bend). (b) Erect penis. Labels: a = glans; b = urethral process (not very long and just sits under the glans); c = strip of connective tissue (raphe); d = edge or ridge of the prepuce.

(a)

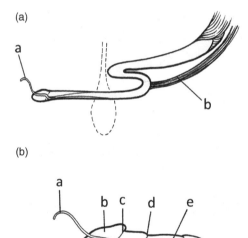

(b)

FIGURE 5.10 Penis of the ram. (a) Non-erect penis. Labels: a = filiform or vermiform append-age (urethral process) (exten-sion of the urethra – semen emerges at its end during ejaculation); b = retractor muscle. (b) Illustrating the ornate end of the penis. Labels: a = urethral process; b = galea (main head of the glans); c = corona; d = tuberculum of the collum glandis; e = collum glandis.

function. In any event, the stallion is a (intra)cervical inseminator and his semen rushes through the cervix directly into the uterus. When coition ends, the mare is said to have been 'covered' by the stallion.

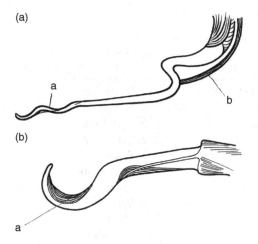

FIGURE 5.11 Penis of the boar. (a) Non-erect penis. Labels: a = corkscrew-like end of the penis; b = retractor muscle. (b) Erect penis. Label: a = ligament attached to the end of the penis, which causes the corkscrew-like action.

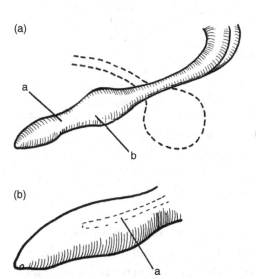

FIGURE 5.12 Penis of the dog (a) Non-erect penis. Labels: a = long (main) part of the shaft of the penis; b = bulbus glandis (area of the penis which swells the most during erection). (b) Erect penis. Label: a = os penis (ossification of much of the cavernous tissue to provide a bone, sometimes called a baculum) (see text).

Equine species are prey animals, so the male dare not tarry too long in the act of coitus, but he will intuitively know that he can probably outstay his predators on a marathon run, so he does not need to worry too much either. The whole act of coitus in horses usually takes about 30 seconds, sometimes a little less, occasionally a little more.

Table 5.1 *Modes of insemination*

Species	Semen deposition	Period of coitus (c) and ejaculation (e) (see text)
Stallion	Intracervical	c: Fairly long
		e: Fairly long
Boar	Intracervical	c: Very long
		e: Very long
Bull	Intravaginal	c: Fairly short
		e: Fairly short
Ram	Intravaginal	c: Short
		e: Very short
Goat	Intravaginal	c: Short
		e: Very short
Dog	Intravaginal*	c: Long
		e: Fairly long
Cat	Intravaginal	c: Fairly short
		e: Short
Rabbit	Intravaginal+	c: Fairly long
		e: Very short
Human	Intravaginal	c: Very long
		e: Very short

*Whole semen passes directly into the uterus.
+Deposited into the anterior vagina (Figure 5.13).

BOARS

Boars yield more massive amounts of semen than stallions, but their coital tactics are quite different from horses and zebras. They have a fibro-elastic penis, and although they are also cervical inseminators, they deposit their semen differently from stallions.

Towards its end, the shaft of a boar's penis is shaped like a corkscrew (Figure 5.11) and it virtually bores its way into the cervix. The first part of a sow's cervix has deep ridges in it, so if you insert a rubber tube shaped like a corkscrew into it, the tube will go in relatively easily if you turn it clockwise (especially with a bit of jiggling), but if you then turn it anticlockwise, it becomes locked in

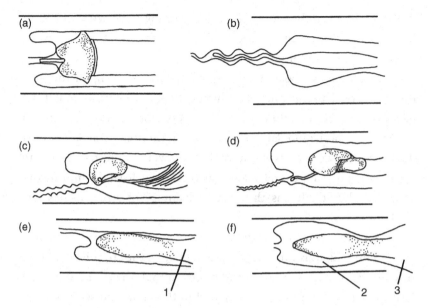

FIGURE 5.13 Positions of penes in coitus. (a) Stallion, intracervical, complete fornix. (b) Boar, a different form of intracervical (no cervical protrusion into the vagina). (c) Bull, intravaginal (virtually no ventral fornix). (d) Ram, also intravaginal (and no ventral fornix), very narrow cervical canal (see text). (e) Dog, intravaginal, but whole semen passes into the uterus (no dorsal fornix). (f) Rabbit, intravaginal. Labels: 1 = bulbus glandis; 2 = front (anterior) chamber of the vagina; 3 = back (posterior) chamber of the vagina.

the orifice and is not easy to withdraw. The boar, therefore, will have a slight withdrawal difficulty until erection subsides. Since the coital act takes 15–20 minutes to complete (ejaculation begins after two or three quick pelvic thrusts and may continue for up to 15 minutes), this is a helpful feature in ensuring that the copulating male stays the course and also helps prevent the backflow of semen. In any case, the gel will bung up the front part of the cervix, so it means that the chance of backflow of semen, which must be a considerable risk with such a large volume, is prevented.

Clearly, coitus in porcine species is long, so what female in the wild or anywhere else would be patient enough to experience another 20 minutes of coitus from another boar, immediately after having had it once? The idea, therefore, that gel is nothing to do with backflow, but rather with preventing other males from inseminating the sow is, to my mind, invalid if not ridiculous.

I have known boars to get so fed up with waiting for the flow of their semen to end that they have fallen asleep mid-ejaculation. However, they can afford to be leisurely, for, as pointed out in Chapter 2, who would dare or care to be confronted by an angry boar disturbed mid-coitus? What animal, moreover, would risk trying to attack a copulating camel, which also takes at least 15 minutes to complete the act. It has the added advantage of size and has few predators anyway.

A sow's uterus has very long uterine horns, but if the organ is excised immediately after coitus and held up in the air, it is seen to be turgidly filled with semen. In the living animal, the seminal plasma is absorbed by the uterus, usually within the hour, and sperms accumulate safely in the utero-tubal junctions. To make sure enough sperms reach this point, sufficient semen must be pumped directly into the uterus. But the semen is not ejaculated in a uniform manner. The first fraction to appear has only a few sperms in it and is mostly fluid; next comes the high sperm fraction with most of the sperms; and this is followed by another low sperm fraction consisting of most of the gel (although some of the gel is apparent in all the fractions). The first fraction, one assumes, provides a watery and nutritious medium for the sperms once they are in the uterus, and the last fraction holds it all in there.

RUMINANTS

All ruminants have a fibro-elastic penis and, as we have seen, are characterized by producing a small volume of semen. It does not contain gel, because backflow is not a problem. The sperms quickly leave the seminal plasma and pass into the cervix, so the seminal plasma is left behind in the vagina. Unlike the situation in boars, therefore, it doesn't matter if the bulk of fluid, which is minimal, is lost by backflow. As already mentioned, passage of sperms into the cervix is made easy by virtue of the cervical orifice being a simple extension of the floor of the vagina. These animals, including not only domestic bovids, but deer, antelopes and other wild ruminants (ovine and caprine animals among them), are all (intra)vaginal inseminators. Bulls usually leap forward with a major thrust when they reach orgasm, so one can envisage the glans bumping into the dorsal fornix, then momentarily rebounding as semen is deposited on the floor of the vagina at the entrance to the cervix. The bull's penis has a ligament on its upper surface which causes the end of the penis to twist somewhat after intromission, but it is much less marked than in the boar. There is also a slight extension of the urethra lying under the main mass (or galea) of the glans, but it too is much less apparent than in the ram (compare Figures 5.9 and 5.10).

The cervix of smaller ruminants such as sheep, goats or deer is very narrow and it is not possible to inseminate these animals artificially through the cervix as is possible with a cow. The procedure has, therefore, to be undertaken by injecting semen into the uterine horns via laparotomy – a small slit in the abdomen allows the horns to be fished out with the fingers. However, it is probably fair to say that artificial insemination in sheep and deer is less common than in cattle, probably largely because the males easily serve so many females. To test which females have been served, a stained block is strapped to the sternum of males and covered in dye. The colour is transferred to the back of the female when she is served, so she can be identified. The contraption is known as a 'raddle'. This technique can also be used in deer (Figure 5.14).

The coital act in large bovine species (such as bulls, buffalos, elands and bison) can take up to 10–15 seconds, but they have less than 5 ml of semen to deliver, so there is no point in prolonging things, even if, with such large animals in the wild, the danger of predators during coitus may be relatively low. This also applies to mountain sheep and goats. By contrast, smaller wild ruminants – such as antelopes, including, in Africa, impalas, springboks, dik-diks, all the gazelles and others, and in Asia, the blackbuck and others, together with even smaller African bovids such as duikers,

Raddle

FIGURE 5.14 Red deer stag with equipment for 'raddling hinds' to test for oestrus.

lechwes and blesboks (but some slightly larger ones too, such as water bucks and kudus) – are all extremely vulnerable to predators, and although feline predators usually have their eye on young ruminants, adults who are otherwise engaged can be targeted. Being caught with your pants down could not only be a matter of embarrassment, but death! So these animals all squirt a small jet of semen into the vagina and get on their way as soon as possible. After all, who cares when there is no issue with seminal backflow and there are so many enticing females to inseminate? Coitus in some of the antelopes may take only a couple of seconds.

DOGS

Dogs are also vaginal inseminators and have a vascular penis. Most of the cavernous spaces in the shaft of the penis have become ossified to form an *os penis* (a bone) or *baculum*. This feature is not only seen in the penis of other carnivores such as bears, walruses, seals and sea-lions, but also of other species, including some bats, small rodents and, in less marked form, some insectivores and non-human primates.

Dog ejaculates are relatively voluminous, but the force of ejaculation seems to drive whole semen (not just the sperms) into the uterus, so that in practice, the dog is really a cervical inseminator, despite its depositing its semen in the vagina. Having no gel in their semen, backflow is prevented by a different mechanism. About two-thirds of the way down the shaft of the penis, there is a bulbous part (bulbus glandis), which swells during erection, reaching its peak size at orgasm (Figure 5.12).

This swelling blocks the vagina, so there can be no backflow. The swelling lasts for about 15 minutes or longer before subsiding, and meanwhile, after ejaculating, the male dog twists around and stands hindquarters-to-hindquarters with the female during this 'cooling off' period. The condition is known as 'the *tie*'. I remember as a child in Yorkshire in the 1930s, seeing women in the back streets throwing cold water over these dogs, presumably in the hope that they would cause the penis to subside, the dog to dismount and a conception be thereby prevented. Personally, I should have thought that it was all too late, but the principle had some vague logic to it. At the time, I have to confess, I did not know what either the women or the dogs were doing. But I was only a little boy!

In the doggy world it is often said that a bitch will only conceive if there is a tie. This is not absolutely true, but the tie is a fair indicator of a 'successful' copulation. From the start of the process up to the point of ejaculation can be up to 1 minute in some dogs, but it is generally rather shorter than this. However, there is quite a protracted period of pelvic thrusting, before orgasm is reached. More, I should say, than in larger animals. The entire procedure, including the tie, can take up to half an hour, sometimes even longer. This is typical of many canine species, including wolves, foxes, jackals, coyotes and so on. Atavistically, we can say that there is no rush, because dogs are descended from pack animals and many contemporary canids hunt in packs, so their alpha males were or are likely to be guarded from predators during copulation by subordinate cohorts in the pack. A bitch is said to be 'lined' by a dog when she copulates, but I am not privy to the origin of this term.

CATS

Both domestic and wild cats have backward-pointing spines towards the end of their vascular penes, as do some of the primates. This may be regarded as their equivalent of a tie, in that it is not easy to withdraw the penis when erect. The fact that some male cats may try perhaps accounts for the screeching of queens from the rooftops in the middle of the night! Such spines occur on the penes of hyenas and spider monkeys, and in these species they fit neatly into corresponding receptacles in the female tract.

THE PENIS: NOT JUST A SIMPLE TUBE

Although ovulation is induced in non-spontaneous ovulators by coitus, it is not yet clear precisely how this happens. It could be argued that there is some substance in the seminal plasma which, when absorbed through the wall of the uterus, enters the blood-stream, and when it reaches the ovary, tells it to release its eggs. However, if this were wholly true, it is likely that ovulation would follow ejaculation by the male much more quickly than it does. As it is, it takes 12 hours in rabbits and up to 25 hours in cats. A more acceptable explanation is a neuroendocrine one, which is to say that messages are sent via nerves to the pituitary gland to release LH. This is supported by the fact that you cannot induce ovulation in a rabbit by simply injecting semen into the doe. In this species, artificial insemination only works if you inject LH intravenously at approximately the same time as semen is deposited.

Since the presence of a male in these induced ovulators normally seems to be necessary, it could be suggested that pheromones bring about ovulation. But the mere proximity of a male is insufficient, because these animals only ovulate if coitus occurs. Thus, one is left with the conclusion that the penis is more than a simple tube for delivering semen and, in these species, it has an additional role of stimulating the vaginal wall, the clitoris or the cervix, which in turn send neural messages to the hypothalamus to release LH from the pituitary.

If this is true, it gives the only reasonable explanation as to why the penis in so many species is sometimes such a very long and ornate organ. The glans of the penis in most mammalian species is a much more complicated object than a simple knob on the end. The glans of stallions extends backwards onto the shaft, whilst ruminants have odd-shaped plates of glans tissue extending similarly. Rams and goats have a twisty appendage, much thicker than a hair, sticking out at the front of the glans and housing the penile urethra (see Figure 5.10). It was suggested at one time that this filiform appendage served to scatter the ejaculate as it came through it and that it was an essential requirement for success in fertilization. In truth, its evolutionary appearance is more arcane, because if surgically removed it has no effect on the fertility of the ram in question. This appendage is to be seen on the penis of a chamois, and a slightly smaller form of it is present on the penis of a giraffe. A wide spectrum of appendages and funny ends to the penis are found in different species of deer and gazelles, whilst the corkscrew shape of a boar's penis is much more obvious in an Indian bovid known as a chevrotain.

There can be differences in the end of the penis even in different species of the same family, such as in rock hyraxes. *Procavia capensis* (the commonest rock hyrax), for instance, has a simple expanded glans to its penis, but a different genus (the yellow-spotted hyrax – *Heterohyrax brucei*) has a distinct short, sharp spike emanating from the glans. Both kinds of hyrax can often be seen in the same group, but each can be identified by their penis! In other species of mammals there are also strange shapes to the end of the penis. In the nine-banded armadillo, for example, there are three separate columns of cavernous tissue, which make the penultimate region of the end of the penis look triangular. The penis of the dugong also has an unusual distribution of cavernous tissue close to its expanded glans, as does the rhinoceros.

FIGURE 5.15 Copulating camels. Well, if it takes so long, why not sit down for it? Courtesy of Lulu Skidmore.

If the mammalian penis is simply a tube for the delivery of semen into the female tract, how has it evolved into such a dazzling structure in so many species? Some would wonder if it is to attract females, but most female quadrupeds never even see the penis, they mostly concentrate on the whole male individual and are only aware of the penis when it is inserted. Moreover, who could imagine a short-sighted female anteater or armadillo gazing admiringly at the penis of a male simply to bring on a period of heat? Yet the penis of the males of both of these species is elaborate. Some other explanation is needed, and it looks as if these features are indeed associated in some way with stimulating the vagina or cervix during coitus.

We do not know whether spontaneous ovulators might not, at some time in their history, have been induced ovulators. There is anecdotal evidence that mares and pigs, for instance, may conceive more easily if they are allowed to copulate rather than be artificially inseminated. It could be particularly important in animals with an extended period of copulation, but much less so in wild ruminants.

You can't get much stimulation from a quick dip of the penis, and it is interesting that artificial insemination in cows is especially effective. But coitus in most non-spontaneous ovulators is, by contrast, a more protracted affair (Figure 5.15).

Our final enquiry into mating includes discussion of how the evolutionary development of an upright posture has influenced the process. For this, it is obviously necessary to look into the question by examining ourselves.

FURTHER READING

Ashdown, R.R. & Smith, J.A. (1969) The anatomy of the corpus cavernosum penis of the bull and its relation to spiral deviation of the penis. *Journal of Anatomy.* **104**, 153–159.

Cave, A.J.E. & Rookmaaker, L.C. (1977) Robert Jacob Gordon's original account of the African black rhinoceros. *Journal of Zoology.* **182**, 137–156 (for students who are interested in rhinoceroses, this is a very interesting historical account of rhino anatomy and is well worth reading).

Cross, B.A. & Glover, T.D. (1958) The hypothalamus and seminal emission. *Journal of Endocrinology.* **16**, 385–395.

Dyce, K.M., Sack, W.O. & Wensing, C.J.G. (1987) *Textbook of Veterinary Anatomy.* W.B. Saunders & Co., Philadelphia, London, Toronto, Montreal, Sydney and Tokyo.

Eliasson, R. & Risley, P. (1968) Adrenergic innervation of the male reproductive ducts of some mammals III: distribution of adrenaline and noradrenaline. *Acta Physiologica Scandinavica.* **71**, 311–319.

Epstein, P. & Zuckerman, S. (1965) Morphology of the reproductive tract. In: *Marshall's Physiology of Reproduction.* Vol. 1. Part 1. Ed.: A.S. Parkes. Longmans Green, London.

Hodson, N. (1964) Role of hypogastric nerves in seminal emission in the rabbit. *Journal of Reproduction and Fertility.* **7**, 113–122.

Hodson, N. (1965) Sympathetic nerves and reproductive organs in the male rabbit. *Journal of Reproduction and Fertility.* **10**, 209–220.

Langley, J.N. & Anderson, H.R. (1895) Innervation of the pelvic and adjoining viscera. *Journal of Physiology.* 19–20 (enormously long papers, but they are classics, so worth dipping into).

Macirone, C. & Walton, A. (1938) Fecundity of male rabbits as determined by 'dummy matings'. *Journal of Agricultural Science.* **28**, 122–134.

Nickel, R., Schummer, A. & Seiferle, E. (1986) *Lehrbuch der Anatomie der Haustiere*. Springer Verlag, Berlin (translated into English by Walter G. Siller), *The Anatomy of the Domestic Anumals*. Paul Parey, New York.

Semans, J.H. & Langworthy, O.R. (1938) Observations on the neurophysiology of sexual function in the male cat. *Journal of Urology*. **40**, 836–846.

Shaffer, N.E., Foley, G.L., Gill, S. & Pope, C.E. (2001) Clinical implications of rhinoceros reproductive tract anatomy and histology. *Journal of Zoo and Wildlife Medicine*. **32**, 31–46.

Short, R.V. (1972) Species differences. In: *Reproduction in Mammals*. Eds: C.R. Austin & R.V. Short. Vol. 4. Cambridge University Press, Cambridge, New York and Melbourne.

Sisson, S., Grossman, J.D. & Getty, R. (1975) *The Anatomy of Domestic Animals*. 5th edition. W.B. Saunders, Philadelphia.

Walton, A. (1968) Copulation and natural insemination. In: *Marshall's Physiology of Reproduction*. Vol. 1. Longmans Green, London.

6 The human male

This is a fairly long chapter, because sexual activity of the human male has dimensions to it that are not seen in the mating of other mammals and it is important that they should be explained. I have to confess to some nervousness in writing about the subject, because I feel to be on shakier ground than when I am dealing with other mammals. However, I trust that what I have written on human mating might be of some interest.

Some authors, even some scientists, have tried to insist that the human male is promiscuous and that the human female displays cryptic choices of her males. This is an opportune argument if you happen to be a promiscuous male, and I do not believe either of these concepts really holds water, but rather that they are convenient flights of fancy for some people. I hope to develop my own line of argument on these issues and to be persuasive. We are bonding animals, otherwise why would we fall in love? And as for human female choice, in the final analysis, there seems little that is cryptic about it to me.

An essential beginning to understanding our mating is a clear appreciation that even if we are proved to be special, we are mammals nevertheless. The redoubtable Oxford don, Logan Pearsall Smith, captured it well, if not entirely accurately, when he wrote:

> This book is written, dear Reader, by a large Carnivorous Mammal,
> belonging to that sub-order of the Animal Kingdom, which
> includes also the Orang Outang, the tusked Gorilla, the Baboon
> with his bright blue and scarlet bottom and the gentle
> Chimpanzee.
>
> *(Pearsall Smith, 1945)*

(As a matter of fact, chimps are not quite as gentle as he thought. They can be extremely vicious.)

Most humans are not seasonal in their mating activity (although Eskimos could be an exception) and females do not have oestrous periods that we can clearly define. However, we have already learned that a lack of seasonality is shared by some other mammals, yet a lack of oestrous periods is peculiarly human. In addition, it is surely the degree of development of our brain that also marks us out as being special. It is questionable whether the size of our brain is all that significant, since Neanderthals had bigger ones than ours (could they perhaps have partly passed them on to us?) and there are other mammals with large brains too. The brain's sophisticated and high-tech development in man, however, makes us very special in all sorts of ways. To begin with, it has endowed us with a unique self-awareness, a desire to worship gods or their equivalent, yet to have a specific extra power of reasoning. We also have a power of imagination that can lead us into all kinds of fantasies. *Indiana Jones, James Bond, The Lord of the Rings, Jurassic Park* and *Harry Potter* all testify to this, not to mention classical literature and poetry and music. By contrast, the drawings of cavemen are more likely to depict the world as it was then seen. However, in our evolutionary history, it is perhaps our amazing skill in communication, even more than tool-making, which seems to have made us stand out from other mammals and even to dominate them. But this specialness superimposes itself on an otherwise basically anoestrous pattern of mammalian sexual behaviour. To understand the importance of this, we first need to ask which parts of our mating style are typical of other non-seasonal mammals.

HOW FAR IS MAN A MERE NON-SEASONAL MAMMAL?

First of all, most human races mate all the year round, so they can indeed be regarded as non-seasonal. Although some men claim they are more libidinous in the springtime, it is not to say that they hibernate in the winter or that they cannot mate in the autumn.

Second, like other mammals, humans indulge in foreplay in advance of coitus, though we shall see that it is a more elaborate process than in species with oestrous periods. Third, human males have the equipment for internal fertilization – a penis, two testicles, bilateral vasa deferentia with ampullae, seminal vesicles, a prostate and bulbourethral glands. They mount their female and intromit with an erect penis, oscillate their pelvis and reach an orgasmic climax, which is accompanied by the ejaculation of 2–4 ml of semen. Semen volumes vary, though I think 13.5 ml is the record. I knew an American student who regularly produced 6 ml every time I asked him for a specimen, but ejaculates like this usually have a rather low sperm count, so the donor yields hardly any more sperms in each sample than anyone else.

Humans are intravaginal inseminators and their penis is vascular. Contrary to the belief of some men, it has no os penis or baculum (no bone in it). Men with good erections may feel that their penis is so stiff when erect that it must have a bone in it, whilst others of lesser potency might wish that it had. But it hasn't. Once intromission has occurred, relatively slow pelvic thrusting is prolonged and may proceed for several minutes, but when orgasm is reached, the ejection of semen is rapid. It is ejaculated in two main fractions: a high sperm fraction and a low sperm fraction. The high sperm fraction is ejaculated first and is followed by the bulk of the seminal plasma, including the gelatinous plug (the clot).

As with all intravaginal inseminators, human seminal plasma remains in the vagina following coitus, whilst the sperms enter the cervix alone. Unlike that of intracervical inseminators or the dog, therefore, human seminal plasma never enters the uterus. In the early days, when this was not always appreciated, clinicians were wont to deliver whole semen into the uterus of women through the cervix during artificial insemination, as if they were inseminating a sow. This often resulted in foreign proteins in the seminal plasma eliciting a reaction in some patients, and it should be no surprise that many of them suffered abdominal cramps after such a procedure. It is essential with intravaginal inseminators that seminal plasma be washed off and the sperms re-suspended in some physiological medium before being allowed to enter the uterine cavity.

So far so good, we seem to be typically mammalian in our copulatory behaviour.

MALE ORGASM

The pleasure of orgasm has never been explained physiologically. There is a pleasure centre in the middle of an area of the brain called the **hippocampus** (the septal area) and it may be that this region is stimulated by repeated impulses passing up the spinal cord from the penis, giving pleasurable sensations which reach their height at the moment of ejaculation. Orgasm varies in intensity in different species. The male rabbit closes his eyes, lets out a cry of ecstasy and keels over with blissful exhaustion. Other species, including most men I guess, react less dramatically. Nevertheless, men are momentarily exhausted after orgasm, their heart continues to beat fast due to circulating adrenaline and sweating is profuse. Extra exhaustion may be due in part to extended pelvic thrusting.

All these events, however, are basically mammalian, so where and how do we differ from other mammals in our mating tactics? The question tempts us to look at our social evolution. It is not easy to assess whether social structures have been and are the main influence on mating styles, or whether the need for certain reproductive strategies has affected social structures. Probably something of both is involved, but it is difficult to discern which is the most important, even in different primates. However, there are both anthropological and anatomical clues to show that we are different from other non-seasonal mammals and have a distinct brand or style of mating, even if we copulate in a similar way to any other species.

ANATOMICAL CLUES TO OUR BACKGROUND

I am unqualified to discuss crucial anthropological evidence in depth, but we can draw some speculative conclusions about our ancestors from human male external genitalia. Relative to body size, human testicles are not very big and not comparable with those of the promiscuous chimpanzee. This suggests that historically and fundamentally, man is not a promiscuous species. It looks as if we were originally descended from mildly polygynous pre-human species, similar to the great gorilla, and that gradually, our human ancestors came to favour monogamy, even if it were serial. Serial monogamy and promiscuity are often confused, but are quite different from each other. Promiscuity involves indiscriminate mating, but serial monogamy includes periods of bonding with a single female, probably until she becomes reproductively incapable. At this point, a more nubile female might be preferred and taken on. This is not an excuse for a 50-year-old man wandering off with his secretary when his wife becomes menopausal – it is simply a matter of such an action probably being in keeping with his biological origins!

If, in a few years, we are capable of predicting when the menopause will be in different women, maybe we shall also be capable of accurately guessing how long a marriage is likely to last! We have no evidence that our male ancestors were totally monogamous, but our sex dimorphism, with men being bigger than women, strongly indicates that they were not promiscuous either. Moreover, there is more anatomical evidence against man being classed as a promiscuous species.

The human foreskin is short and only covers the glans (though it varies in length in different individuals), whilst the whole penis protrudes from the pubic area, but hangs loose when not erect. Thus, the human penis is a particularly obvious organ. Framed as it is in darkish pubic hairs, it is clearly visible from afar in a naked man, even though his facial features might at that distance be blurred. It is possible, then, that the large and conspicuous human penis could be construed as a product of intersexual selection, an adornment for attracting females like a peacock's feathers. But, if we run true to mammalian form, it is also likely, and probably more so, to be an instrument for announcing one's presence to surrounding males.

It could have been important for our forebears to know whether that figure standing on the rock or swinging from the boughs was male or female, because competing males could take it as a warning if it was recognized as being male. But for whatever reason, most men hope that their penis matches up to or surpasses that of competing males when it comes to size.

But what if the colour of your skin is black? Anecdotally, it has been pointed out that the non-erect penis of men of dark skin colour is larger than that of most White or Oriental men, so this could compensate for it being less visible by virtue of colour contrast. It is interesting to note that the penis of Asian Indian and Pakistani men is darker than their body colour, so this should help also to make it stand out at a distance.

A large and protuberant penis might also serve to cement a bond, and if it is argued that it is more satisfying to a female than a small one, the larger it is the firmer the bond is likely to be. Unfortunately, discrepancies in the size (especially the length) of an erect human penis are by no means firmly established, if indeed they occur at all in a significant sense, so this suggestion seems to be a rather weak one. But it is surely likely that a conspicuous flaccid human penis has been of some special advantage to man in the past. It certainly provides further evidence that man is not a promiscuous species. The non-erect penis of the promiscuous chimpanzee, for example, is most unobtrusive (in marked contrast, incidentally, to when it is erect). You see, the chimp has no need to announce his presence, he just gets on with things, before others in the group do it for him.

That humans have no very clearly delineated sperm store in the tail of the epididymis indicates that there is never a sudden or continuous demand for a profusion of sperms, as there would be if we were seasonal maters or were fundamentally promiscuous (see Figure 3.21). It seems rather that in man, sperms are being slowly but constantly produced. A daily demand is unusual, even if some coital episodes may involve two ejaculations. But if intercourse is too frequent, there is a rapid decline in the number of sperms ejaculated. For this reason, when semen is produced for analysis, a 2–3 day abstinence period is requested for standardization. The emission reflex easily and quickly tires, and in a species with a diminutive sperm store such as man, this serves as a good mechanism for ensuring that demand never exceeds supply. Clearly, though, after a long period (and remember that, compared with other mammals, men live a long time) the total number of ejaculations in men is high, but it is not the case over a short period. Men can, therefore, be regarded biologically as rather lazy maters who do not need a high sperm output during any given restricted period. As we have seen already, gorillas are even lazier.

If it be accepted that man is truly monogamous, what about those men who are polygamous? Polygamy is clearly a cultural issue, but it involves bonding and is nothing to do with promiscuity. In some communities, multi-monogamy and multi-bonding (polygamy) appear to be unremarkable, although some tribes do indulge in a degree of polygyny.

WHAT CAN WE LEARN ABOUT OURSELVES FROM OTHER PRIMATES?

In terms of our mating habits, the answer to this question seems to be 'very little'. First, data on other primates are confusing, to some extent, because many observations have been made on captive animals and it is known that captivity can profoundly affect mating behaviour. Observations in the wild are probably more valuable. Saying this, however, the environment and other parameters can be controlled in captivity and this is extremely important in research. But are there any likenesses?

Menstruation is a human characteristic, so does it occur in other primates? Yes it does, but as we have already learned, it is typically confined to Old World monkeys, although there is a hint of it in other primates such as the tree shrew and others named later. Old World monkeys are not seasonal in their mating activity either, but most primates, including New World monkeys, are, even though it varies in different species. One species of loris (a prosimian – i.e. non-monkey – primate) breeds only twice per year, whilst another breeds all the year round, and little appears to be known about the mating activity of pottos (a type of loris). The Malayan tarsier (another prosimian) breeds throughout the year, and probably the tarsiers of the Philippines do too. But yet other prosimian primates, such as most of the lemurs of Madagascar, have a single distinct mating season. But even among lemurs, there is an exception, in that one species has two mating seasons.

Among the anthropoid New World monkeys, the squirrel monkey and the titi monkeys of Brazil can breed throughout the year (although in captivity the squirrel monkey has been reported as being distinctly seasonal). Most of these animals have only a single off-spring at a time. Cebus monkeys, or capuchins and spider monkeys, unusually for New World monkeys, have a menstrual cycle and breed all the year round. So there are some physiological similarities to humans in some species. Then there are the owl monkeys or dour-acoulis of South America, which are nocturnal and presumably mate at night – very coy! Like humans, though, they probably mate throughout the year.

Among these New World monkeys, it could be said that marmosets and tamarins are the most like us in that they bond to the extent that the male is involved in the rearing of young. They usually have twins or triplets and both of the bonding sexes are the dominant ones in a group. Others in the group may stray into other groups and can return without any animosity. But, in spite of there being many relatives within a group, there is quite a bit of intergroup aggression. Groups mix with each other amicably, but group members will join together fiercely to fend off any individual who tries to interfere with a bonded pair in their group. It is all rather like an extended human family or clan.

Some New World monkeys indicate that they are in oestrus by swelling of the lips of the vulva, whilst others show no sign that is visible to the human eye. Because of this social and sexual diversity between species, New World monkeys tell us little about ourselves. They do show, however, that although they employ a k-strategy of reproduction, social structures play a more important part in their sexual behaviour than either proximate or ultimate principles. These monkeys also demonstrate a tendency for females to attract males by showing that they are in heat, rather than leaving the males to find out for themselves.

If we turn to Old World monkeys, we can glean a little more about our own sexual behaviour. In the first place, all Old World monkeys menstruate, even though typically they have a distinct period of desire or oestrus in between the menstrual periods. Once more, this underscores the difference between oestrus and menstruation. The length of oestrus differs in different species, but clearly, ovulation occurs between the menses. After ovulation, the inside of the uterus (endometrium) is built up and thickened in preparation for an anticipated pregnancy. This is known as the 'luteal phase' of the cycle. If pregnancy does not supervene, there is a massive breakdown of endometrial tissue – hence the bleeding. Menstruation, therefore, is simply a sign of a disappointed uterus. This applies to man as much as to Old World monkeys and some other primate species.

In this regard, our reproductive cycle and that of our relatives the great apes are similar, except for women having no period of oestrus. So what about their social behaviour? Is there anything we can learn about ourselves from this? A definite answer does not appear to be possible at this stage, because the socio-sexual behaviour of different species of Old World monkeys is so variable, it is hard to pick up threads that they have in common. For instance, it is not safe to try and equate the socio-sexual situation of lowland gorillas with those of mountain gorillas. Which primates are seasonal and which are not is not easy to determine either, because animals that may be seasonal in the wild can lose their seasonality in captivity, where feeding and other conditions are better. So the subject of primate reproduction generally is complex and variable, and we still have a great deal more to learn about it.

A feature that does seem to emerge is that, in the wild, mating primate males are usually the dominant (alpha) males. Moreover, if they are not of a promiscuous species, they are larger than their female counterparts. This fits in nicely with our own sex dimorphism. But if we put social factors to one side, what are the sexual signals that pass between male and female to draw them together?

SEXUAL SIGNALS BETWEEN NON-HUMAN PRIMATES

Here, again, we have a wide spectrum of variants. However, overall, it is fair to say that non-human female primates send out much stronger signals that they are ready to mate than most other female mammals. This is strikingly demonstrated by Old World monkeys, especially perhaps by the macaques, such as the rhesus monkey, and also by chimps. These animals, like the baboons, have an area of specialized sexual skin extending variably in different species, but generally occurring around the anus and extending to the vulva. The base of the soft cleft between the anus and the vulva is called the *perineum* and this kind of sexual skin occurs at each side of the perineum. After the menses, there is a short phase during which this sexual skin does not change, but when the animal comes into oestrus, the area swells and takes on bright colours of blue and, especially, red hues (as observed by Logan Pearsall Smith, although, because of its swelling, it can hardly be missed!). It then subsides just before the next menstruation.

This is like turning the fairy lights on to tell the male that you are ready to go. Indeed, female primates have been seen to stick their enlarged and lurid backsides under the nose of an eager male, just to make sure he knows! The males also have sexual skin, which extends towards and then over the scrotum. It does not swell to the same extent as in the female, but in both sexes it is governed by steroid hormones, oestrogen in the female and testosterone in the male. The Galada baboon, which spends a great deal of time sitting upright on its bottom, has a heart-shaped area of red sexual skin on its chest instead.

Unlike most quadrupeds, therefore, apes and other Old World monkeys show a tendency towards mutual soliciting. Male chimps display before their females with an erect penis of startling length, but their female targets only come running if they are in heat. The male orang utan gives a clarion call to announce his presence, but when the females are receptive, they put on the come-hither act as well. The strange ethereal call of a male gibbon doubtless betrays a hopeful cry also, before listening for a female response.

There can be no doubt that among primates, especially Old World monkeys, females in oestrus are pro-active when it comes to luring males into having sex with them, and the males let them know through sexual skin and perhaps an erect penis that 'Barkis is willin''. This mutual display as a pre-coital tactic is quite different from most other mammals. It doubtless has a social basis and is more familiar to other primates such as ourselves than adopting a technique of simply going on to the main course and missing out the starter, as other mammals do!

Furthermore, the mutual handling of genital organs before coitus has been commonly observed in non-human primates, as well as *cunnilingus*, although *fellatio* seems to be distinctly human.

PHYSICAL SEXUAL SIGNALS

These brilliantly coloured bottoms show us that although most primates have highly developed forelegs, mostly as arms, they are fundamentally quadrupedal, although sometimes with the rear end higher in the air than the forequarters. Since many species are arboreal, this is especially true as they descend a sweeping bough. It is, therefore, the back end of the animal that sends out the signals for male attraction. This applies to all quadrupeds. A cow, for example, might show streaks of blood on her vulva when she is in heat, and swollen vulvae are common among animals in oestrus.

However, the human primate has acquired an upright posture (*Homo erectus*), so the bottom is less prominent, meaning other physical devices for attracting males have had to evolve. It is evident

FIGURE 6.1 Physical sexual signals in chimpanzees (and all basically four-legged primates) are at the rear end. © Photolibrary.com

that everything in this regard has slipped round to the front. A female chimp's backside may be a thrill for the males, but it is breasts and the so-called 'mount of love' (the rounded part of the pubic area above the vulva) that are a turn-on for most human males. This is not to mention the important role played by the face.

There are men who have a particular penchant for the callipygian (shapely buttocks), and this is of interest because breasts are definitely natiform (soft, rounded flesh with a cleavage in it). Breasts, for such men, may come across as simply being buttocks with headlights on them. The headlights are important too, because the areolae around the nipples are all part of the attraction. In spite of this slipping round to the front, the women of some East Indian and South American tribes show varying degrees of steatopygy (large fat deposits on the rump), which presumably is in part a lure for their men. So, it looks as if the backside has not totally lost its appeal for human males. We may note that Velasquez's *Rokeby Venus* highlights the buttocks rather than the face or breasts.

It should not be forgotten, though, that the development of the gluteal muscles in man relative to the hamstrings is primarily associated with the upright posture rather than sex, and cannot be used by quadrupeds for sexual attraction. Perhaps this is why some quadrupedal primates have developed sexual skin instead.

Copulation in most mammals involves contact by the front of the male and the back of the female, although pygmy chimps and orang utans usually do it front-to-front. But a larger male chimp, although he may show-off to females by sitting down and waving his phallus in the air, copulates front-to-back. Because of his upright posture, the human male usually approaches his female from the front when he is about to copulate, since it is primarily the front of the female that attracts. During the actual act of copulation, it has become quite fashionable for the man to roll over onto his back and let the female do all the work, but this can be little more than acquiescence to a female partner who wishes to feel dominant. Copulation front-to-back might have been common in the past, when our ancestors were essentially quadrupedal and is also popular with some contemporary couples. However, this approach is probably not the rule, even though the large human penis allows coitus to occur in all sorts of positions, as well demonstrated in the *Kama Sutra*. It is likely, although this can only be speculative, that the 'Christian' (front-to-front) position remains the most usual human practice.

As far as we can be aware, the human male is the only one that is attracted by mammary glands, but apart from those of a dairy cow, other mammalian mammary glands are much less conspicuous than those of women, and are hardly apparent at all in a non-lactating animal. If it is the natiform nature of mammary glands that men find erotic, then they should be able to be callipygianists and yet be mesmerized by breasts with equal facility. When, therefore, a lascivious and uncontrolled male pinches or strokes a lady's derriere, it presumably gives him a similar sensation to squeezing or touching her breasts. But it is apparent that the upright posture has had a profound influence on the parts of the female body that attract human males.

There are other physical features that electrify the male hypothalamus, but these are specific to an individual. Facial features, especially eyes and lips, play a part (again associated with the upright posture), and the colour and texture of hair are important to some men, even including the way in which a woman arranges her hair. Shapely legs, and often long legs, in a woman are desirable to most men, which is also probably associated with bipedality. But many of these attractants are first viewed from the front when we greet each other. Such pre-copulatory frontal greeting takes place in other primates too, and perhaps this is why chimps, gorillas, orang utans and baboons, who spend quite a bit of time sitting upright, have also developed facial expression.

Like other female mammals, physical features seem to be less important to women, as regards sexual attraction, than they are to men. There may be a biological explanation for this. If a man is to copulate, he first needs to bond, but he is only likely to do so if he thinks his female is likely to agree to it. Since, however, she has no period of heat, the male needs to look for other signals of receptivity, and a nubile woman usually indicates a potential for receptiveness in all sorts of ways, rather than a women who is careless of her appearance or gives the wrong facial or other signals. Women, by contrast, seem to look primarily for a man who, figuratively speaking, might be likely to help in nest building and be of a loyal and caring nature. If they find these qualities, they are more likely to bond. This is probably important, because fine physique in a man certainly does not reflect his fertility. A funny little chap can be quite fertile and his offspring could be bigger and physically better than he.

This is not to say that physical features play no part in female attraction, but a man's potential for offering security and loving tenderness probably play a greater role. This must surely be why beautiful young women are sometimes seen to be gazing adoringly into the eyes of an ageing millionaire. Some might think it is a good motive for becoming a millionaire! But I have heard several women speak of a variety of male physical features as being attractive to them, if perhaps less compulsively than female features are to men.

One lady told me that the back of a man's head was erotic to her, because it had a connotation of vulnerability rather than aggression. Judging by passing comments, it seems that well-shaped muscular buttocks in a man are attractive to lots of young women (Dr Desmond Morris tells us that this is because they indicate good thrusting power). Large pectorals in a man denote strength, and an overall good physique indicates potential dominance over other males, if that is important to a searching lady. Certainly, puny, concave-chested men who do not fill their trousers are hardly erotic. From what I have heard from many ladies, however, really huge muscles are abhorrent to most women. Although facial features can also be extremely important to women, overall countenance, demeanour and a good sense of humour usually seem to have priority.

A number of women find it attractive if a man sports chest hair at the top of an open-necked shirt, although now, I am told, hairless chests are in vogue and hirsutism is out. It all goes to show that sex can have its fashions.

An exposed navel appears to be an erotic attraction to both sexes, which brings us to the uniquely human quality of imagination in sexual activity. For some reason, an exposed navel can elicit thoughts of coitus, and underlines the essential role of imagination in human erotica. The partially clothed body is widely recognized as being more erotic than the totally naked body, especially for women, unless, of course the observer is in a state of sexual arousal. But a man in swimming gear, for example, exposes his navel and, again, it must be significant that the navel, a washboard abdomen and phallic bulge are all at the front of the body, as also are the rounded abdomen and exposed navel in women. We have already seen that anticipation can be a powerful sexual stimulus in men, and I have heard it said that for some men, imagined sex – the fantasy – can be better than the actuality – that is, the real thing. Doubtless this often applies even more to women.

SMELL

Smell is not without importance in human sexual attraction as well. Some women find fresh underarm and body sweat magnetic, but unfortunately it soon goes stale. Even in a freshly washed male, scrotal sweat has its own special smell, as does that of the male perineum, just as the smell of vaginal secretion is unmistakable. Probably, these odours are all vaguely related to pheromonal attraction and are mere leftovers of a smellier past.

Human semen also has a distinctive, tarry smell. This is due to a substance called *spermine*. It is an organic base put out by the prostate, and it tempts the thought that perhaps our forebears might have been territorial and marked out their territory rather like a dog or a wolf. Human semen is the only semen we can smell – that of all other mammals is quite odourless to us, unless it is contaminated with urine. This suggests that only the sexual smells of our own species are recognizable to us, and there is some evidence that human pheromonal smell may be specific even to individuals. In modern parlance, each man has his own chemicals or MHCs (male human chemicals) in his sweat and it is possible that these can be picked up and prioritized by the opposite sex.

The evidence seems to be, though, that quadrupedal mammals have a more highly developed sense of sexual smell than most humans. It is a common observation that dogs do an awful lot of sniffing around, and even other species may be the target there. Do many dogs not invariably make a beeline for the human male crotch? This could be an instinctive attempt to suss out whether the guy is a dominant male. Horses have a special area in the nose (called the *vomeronasal organ*) that consists of modified olfactory (smelling) tissue. When stallions have a serious sniff, they throw the smell back onto their nostrils by curling their upper lip and taking a deep breath. This action is called *flehmen* and is often used by stallions in determining whether or not a mare is truly in heat. Other animals with a long upper lip occasionally do it if they are capable, but the vomeronasal organ is essentially equine.

FIGURE 6.2 A stallion displaying 'flehmen'. A man cannot sniff like this!
© Photolibrary.com

OTHER SIGNALS

We have seen that although the great apes menstruate, they have a recognizable period of oestrus between the menses, so the human female is essentially unique among mammals in having no period of heat at all. This being the case, how can a human male know when his female is receptive – that is, when she will allow him to copulate? Here we get into deep water, because of our being a bonding species. The first signals, therefore, must be that each partner wishes to bond rather than immediately to copulate. These signals may be very subtle and include a variety of mannerisms, sometimes unconscious ones, facial expressions (is there not that glance, that particular look?) and body language. The signals can involve language itself, too, verbal hints of a wish to bond, for example.

But what can be the advantage to man of the evolution of bonding? Here, again, we meet with the reality that human females have no periods of oestrus. Preparation for a bond (protracted courting) enables a male to determine whether or not he has chosen a female who has similar feelings of attraction. It might be said that there are easier ways than prolonged courting to find out, but we are talking here about how things might have evolved. We might remind ourselves that our ancestors were likely to be less articulate than we are today, and the formation of a protracted bond could enable a male better to recognize, through familiarity, when his female was receptive. This is not to say how bonding evolved, because some mammals with oestrous periods also bond. Bonding, however, must be useful in learning to detect receptivity if your female never comes into heat – and it is all to the good that human females don't. Can you imagine the mayhem that would result if they did!

Men do not go in for flamboyant short bouts of courting like birds (which also display no oestrus), so we see that among such anoestrous animals, different techniques are used by different species for testing out and ultimately seducing their females. But none of the techniques are necessary in species in which females display regular periods of oestrus. In this regard, therefore, human sexual behaviour is quite different from, and thus bears no comparison with, that of other mammals. This is in spite of their being equipped with similar organs for actual copulation. The lack of oestrus also means that pairing and reproducing in humans is more of a hit-and-miss affair than in other mammals, a quality which could lead to a broader genetic mix and account for extreme diversity in human populations. Animals that have periods of heat are impelled to mate with an active male in their vicinity, but this is not so with a human female, so if a man realizes he has made a mistake, he will simply go off and try again elsewhere. He might be successful next time.

Physical behavioural signals soon follow the earliest signs of attraction – perhaps a touch or a kiss. A kiss on the cheek is usually a greeting or a token of felicity in the Western world and may be merely a sign of respect or affection. A kiss on both cheeks is fashionable ('I have been to France!'), but the insincerity of an air kiss is, to my mind, irritating ('I don't wish to disturb my make-up or to make any

FIGURE 6.3 A fond farewell or an invitation to bond? Behavioural sexual signals.

physical contact!'. So why not bow or shake hands?). By contrast, a prolonged and serious kiss on the lips is usually taken as an invitation to bond, but even that is sometimes difficult to recognize accurately. Usually, though, it indicates that the man who gives a kiss on the lips has some feeling for the lady being wooed or about to be wooed. Particularly passionate kisses, which might even include lingual exploration of the tonsils, are surely an invitation to copulate rather than to bond, but we are at the attraction stage here, so gentler invitations to bond are more likely.

Flirting is probably a light-hearted game of indicating a willingness to copulate. But when a man approaches a woman, probably when he is drunk, and says 'I would like to go to bed with you' almost before he has had the courtesy to say 'good afternoon', he is probably more serious than a flirt and is not being true to his species. He might get a one-night stand out of it, but then we have seen that one-night stands are not really characteristic of the species either.

Common observation suggests that courtship displays by other primates are usually invitations to copulate, whereas in humans they are more accurately construed as a preparation for bonding, with each sex testing out the other as a possible future partner. But potentially good copulatory ability is not the only parameter being sought. Possible partners might instinctively or consciously judge genetic potential for all sorts of different qualities. For instance, they might be looking for likeability, generosity, kindness, a good sense of humour, common interests and, in some cases, brain power. These are all included in the assessment. They are very important for potential bonding and there is some urgency in the quest, because human bonding is meant to be for life.

So human courtship is normally quite prolonged and might, in some cases, end in dithering because, if the bond is to last, each partner needs to become a close friend of the other and to experience a genuine blending of personality. There may be differences of personality and certainly of opinions, but there needs to be some degree of synchronization in this regard, some compatibility, because it is difficult for a meaningful bond to be made with someone who constantly irritates you, even if she is beautiful or he is handsome! In this respect, humans are helped in their bonding by falling in love.

This whole rigmarole is part of the mating process in man, but fundamental human behaviour patterns are such that if a man is to be true to his species, bonding with his female has to happen before copulation takes place. This is how we have evolved. It is for this reason that, unlike most other mammals, humans naturally have an extended period of courtship as a preamble to coitus. Unlike some birds and fishes, this period is not an invitation to copulate through sexual display, but rather a time to reflect on whether or not to form a permanent bond. Even when a bond is formed and coitus beckons, human foreplay is exceptionally long by mammalian standards, because there are no distinctive or immediate physical signs of female receptivity. There may be behavioural signs, but they may not be compulsive, so an extended physical preamble (foreplay) is needed before a human male can be absolutely sure.

FALLING IN LOVE

Why people fall in love with one person and not another is not easy to explain. We cannot be certain what cues release the feel-good hormones, serotonin and dopamine, and what actually happens to the hypothalamus in the process. This has tempted the suggestion that for an initial attraction, all sorts of mysterious factors are called into play, such as particular and individual pheromonal chemicals. To my mind, the evidence for such factors is flimsy, although it is not totally impossible that they play a part. Other suggestions of cryptic female choice and wild speculations about hidden sex appeal are, to me, little better than astrology. Personally, I applaud instinct in making decisions, because I believe that it is based on stored evidence. But this does not make it exactly cryptic, because we might find out – eventually – how it works.

We have seen that falling in love is a major asset in helping us to bond, and the physiological basis, or something similar to it, may occur in all bonding species, including the otter and the gibbon. There is some evidence that when someone falls in love, the back part of the pituitary (the neurohypophysis, see Figure 3.4b) is stimulated to release its hormonal component (*oxytocin/vasopressin*) and that this may be important. Nerves pass from the hypothalamus to this part of the pituitary, where the hormone is released. The cue for oxytocin production and release via the hypothalamus is unknown, but its functions in the female are well recognized. A role for oxytocin in the male has been rather vainly sought for several years, so its possible participation in love is interesting. Whatever the detailed truth may be, we can confidently say, although it might not seem very romantic, that love is very much a hypothalamic affair.

What marks us out as being different from other bonding mammals may simply be that we are aware of being in love, that curious compulsion to enjoy the constant company of a particular individual of the opposite sex. In other words, love is a physiologically based drive to bond. The system may kick in early in a relationship (love at first sight), but more often, enduring love needs time to develop. This might be why bonding normally needs to be prolonged, but could, paradoxically, explain why pre-arranged marriages often turn out to be successful. In any event, it is apparent that whatever the cultural mores might be, human mating is a sophisticated business necessitated by an anoestrous state and a basic biological need to bond.

MARRIAGE

I am unaware of any human culture that has no religion of any sort. This suggests that a need for some sort of worship is an innate human quality. In Russia the orthodox church sprang back into prominence as soon as the enforced secularism of the Soviet authorities went away, and there is hearsay evidence that it functioned covertly before that. When churches and religions have been persecuted, history tells us that, in many instances, they survive. It is not surprising, therefore, that acknowledgement and confirmation of the bonding of a man and a woman usually involves a religious ceremony. It is interesting that so often when couples who have lived together outside marriage finally decide to marry (usually when they are ready to have children), they so frequently choose to do so in a church. It never seems to occur to them that, if they have copulated before bonding was assured, they have broken all the rules of the Church, not to mention the biology of their species, unless, that is, copulation can nowadays be regarded as part of the bonding process. However, it does suggest that the religious ceremony of marriage is quite important in human communities, perhaps especially to women, whose monogamy, as already explained, is less likely to be serial than men's. A church ceremony is not a matter of dressing up either, because they can do that in a registry office or in someone's garden.

There is usually a distinct reason for non-religious marriages; a wish for church marriages is probably culturally or tribally based (pressure from parents and family?). It applies, perhaps especially, to religions other than Christianity. The fact that it is traditional to marry in a church also comes into the picture, of course, but a spiritual or even metaphysical dimension seems to be involved in the custom.

Although a few years ago a 25 per cent reduction in church marriages was reported in British newspapers and is perhaps a signal of increased secularism worldwide, in typically human fashion, religious fervour remains undiminished. The significance of humans wishing to worship a god or gods is not relevant to our discussion here, except in so far as that it may influence human mating behaviour.

Up to this point, discussion of marriage has been centred mainly around Western culture, but it applies to most human cultures around the world in one form or another. Courting has no chance to operate in those cultures where marriage takes place before each partner has even seen the other. All that can be said about that custom is that it is contrary to the evolutionary processes we have discussed, and thus to human instinct. But if it works well, so be it. These customs simply mean that a marriage occurs before bonding, rather than it being a token that a bond has already been established. It passes over an awful lot of fun, but then perhaps fun is not a key factor in the cultures or religions involved. The principle, it seems, is simply that in some customs, bonding can only occur after a marriage ceremony has been enacted.

UNCONVENTIONAL SEXUAL BEHAVIOUR

As startlingly revealed by the text of Malinowski in his *Sexual Life of Savages* and by other scholarly works, different cultures have rites and rituals concerning sexual behaviour that might appear to be peculiar to the Western eye. Even in Western society, many people obtain sexual gratification other than by our conventional mode of mating. But we are primates, even if we are rather special ones, and our forebears were primates. One only has to visit a zoo to witness what are, to us, some strange goings on among the monkeys and apes. Chimpanzees can be seen watching each other copulate, and age appears to be no barrier to the exercise. Admittedly, the onlooker usually seems to be rather bored, but it nevertheless occurs. We might ask, therefore, if similar voyeuristic shenanigans among human populations might simply be an evolutionary left-over from our fore-bears, or whether it is just an individual's perversion of sexual desire. We have already seen, though, that we shouldn't read too much into monkey behaviour when it comes to trying to interpret our own.

Thus, when human sexual behaviour is eccentric, it is some-times difficult to decide whether it should be construed as normal (accidentally inherited by some people from the past and simply representing the boundaries of a wide spectrum of sexual behaviour patterns) or whether it should be recognized as being abnormal (a distinct psychological disorder). Night-clubbing leading to instant coitus, group sex, bondage and gratification through watching porn-ography are all undoubtedly sexual, but may be more about having fun and abandoning restraint than about the real process of mating. Yet, many such essentially extra-marital activities (even if sometimes indulged in by married couples also) are condemned by different soci-eties as being depraved. It all depends on the definition of depravity, which varies between different groups, cultures and religions. Accept-ability also changes within groups with the passage of time. Not long ago, there was an unmistakable distaste for masturbation throughout the Western world ('You'll go blind!'), but today it is mostly tolerated, perhaps on the grounds that everyone knows it goes on.

In most cases male masturbation seems to be no more than a release from a prolonged period of abstinence in circumstances of there being no girls around to woo. Even if there are, it might be safer than going on the rampage. It has even been recommended for the relief of tension. Such autosexuality can become habitual, especially if a touch of narcissism is involved, but who cares? It is an entirely harmless exercise if undertaken in private – even bulls, rams and dogs do it from time to time. It happens less when the real thing becomes available. Sado-masochism is less well received, but we now have a society in the West that worries less than it used to about people acting out their sexual urges, provided they are performed consensu-ally and privately. This trend is regarded by many, not only those in religious groups, as a dangerous societal development. We shall see.

But even in Western society, with its liberal attitudes, we would still be less sanguine than would a chimpanzee, if we were to witness a naked couple copulating in the main street.

All this teaches us that cultural and religious values have been superimposed on the sexual delights of many human beings, whose sexual imperatives might or might not be typically primate. Such man-made rules for governing acceptability have complicated the issue and mostly seem to have their origin in religion. Both St Augustine and Thomas Aquinas held that sex in itself was sinful, so they have an awful lot to answer for in terms of guilt complexes. They have certainly muddied the waters, and in this essentially secular Western society of ours, if not in the world at large, it is perhaps timely to reassess what sexual behaviour can be taken as acceptable, what can be tolerated (turning a blind eye, even if the activity is eccentric) and what, if anything, is to be deprecated and discouraged. The rules need to be looked at again, even if a consensus might be extremely difficult to achieve, the more so since society itself evolves and changes its attitudes.

Throughout the nineteenth and early twentieth centuries, boundaries of acceptable sexual behaviour were relatively clear throughout the Western world. In the 1960s, however, society moved the goal posts so that nowadays, almost anything goes for some people, provided it is not done in the streets and frightens the horses! But individual views still differ quite widely, in spite of self-assertive opinion-makers frenetically telling us what we ought to think.

Take public nudity as an example. It is widely frowned upon in the West, but my feeling is that when nineteenth-century missionaries insisted on savages wearing clothes, they were being extremely arrogant in believing their culture to be superior and thus worthy of being imposed on others. As far as I can see, we only started to wear clothes in order to keep warm or to protect the genitals, not for purposes of modesty. I am sure fundamentalist Christians would disagree with this view, so reactions to nudity are ambiguous. 'Skinny dipping' (swimming in the nude) in front of other people, 'streaking' (dashing stark naked across a football field or elsewhere) and 'mooning' (bearing your bottom) are all likely to result only in nervous laughter by onlookers or a condescending smile (even if it results in a police arrest!).

These sorts of actions may be intended as a joke, but are fundamentally sexual in connotation and are often deliberately intended to shock. This does not apply to naturists. But although a figurative blind eye may be turned on public nudity, we find genital exposure in a fully clothed man unacceptable. It could be argued that streaking and the like are just a lark, whereas a man exposing himself is more sinister and unpleasant. Nevertheless, this surely tells us that our modern senses of sexual morality are more confused than rational. It also shows, however, that instinctive, if diverse, human sexual behaviour is very much constrained by cultural rules. Infidelity, for example, is generally held to be an absolute 'no-no'.

INFIDELITY

We all know that there are some men and women who, in spite of having bonded either in marriage or a long-term partnership, occasionally have a 'bit on the side' or even indulge in one or more full-blown affairs. It is not sensible to try and compare this with similar activity among monogamous birds, since they do it for reproductive insurance. Each unfaithful human partner is not usually concerned about that and they may have a fling for a variety of different reasons. But to try and read some hidden biological cause into infidelity, beyond straight forward serial monogamy, is, to my mind, unnecessary and is probably a fantasy. Worse still, to suggest the selection of semen samples by human females can only be guesswork and is devoid of serious evidence. I emphasize again – we are highly successful mammals, not birds.

The fracture of a bond through female infidelity, I would argue, is more likely in humans than in other bonding species of mammal, because the female has no periods of heat (oestrus) and is capable of conscious choice of a mate. MHCs might be a factor in selecting that mate, but I think that money and its attendant opportunity for a good time (perhaps not available from an impecunious husband) could also be a temptation! Is it not so, that after a flutter, many women go back to their husband? By contrast, the heritage of males being one of serial monogamy might impel some men to veer off into an occasional dalliance. But men, too, often try to take up with their original partner again. Provided it is not too widespread, these bouts of infidelity are of little consequence to a society, however much they might affect a bond. However, this is doubtless a reason for marriage ceremonies in all religions being shrouded in vows of faithfulness, because most of our civilizations are built on a fabric of pair bonding and the production of children.

HOMOSEXUALITY

The subject of homosexuality is too complicated to discuss here in depth, but since it has become an issue of some topicality, perhaps it might be worthwhile trying to look at it from a biological standpoint. I must warn the reader that this section and the following, cover issues that some readers might perceive as sensitive. I would like to emphasize that the intention is not to offend, only to present the facts and lay the case open for debate.

Although males of other than human mammalian species can occasionally be seen to mount other males, it does not usually end in intromission and is most common in neopubertal males. It might occur if rams are penned together for a period without any ewes, and it has also been reported anecdotally in humans, that seafaring men might take a male partner whilst at sea, but plunge into a heterosexual brothel immediately upon reaching land. Such homosexual encounters can only be construed as deprivation (or desperation!) and probably do not reflect true homosexuality. Lust can be indiscriminate. Remembering the graded preferences of rabbits, for some men who seek sexual relief, it might be better a human body, even a male one, than none at all. A society does not even need to acknowledge the existence of such acts, provided they are undertaken consensually and in private.

Similarly, mutual masturbation is not uncommon among heterosexual schoolboys, who are simply experimenting with sex. I have been told that there are also individuals who habitually indulge in homosexual acts, be they homosexual or heterosexual men, and such behaviour may be nothing more than slipping over into promiscuity and behaving like a chimpanzee. But this is a moral issue because, when it comes to such lustful behaviour, there is a choice and the availability of restraint or refusal on the part of any individual.

It seems to me that bonding homosexuals fall into an entirely different category from those who go in for casual homosexual sex, because they have no choice and, depending on your moral values, a compulsive homosexual partnership could be interpreted as being strictly moral, because it is certainly a form of mating. Perhaps the term 'homophilic' might be more appropriate. By contrast, since casual homosexual sex is a matter of acquiescence or refusal, it is not essentially different from any other expression of lust. But if a man is driven to pair bond with his own sex after falling in love, it must surely be a case of mistaken sex identity. That this is a physiological or psychological disorder is an inescapable conclusion, even if it is not particularly important. With luck, we shall be able to explain its origins one day, but it might take some time, because, since we are not dealing with a socially or individually serious condition, funding for research is likely to be in short supply. Bisexuality is even more difficult to explain – when a man is capable of falling in love with either male or female, he reveals what a complicated subject this really is.

Estimating the incidence of true homosexuality is extremely difficult, and although it is probably less common than some activists would have us believe, it is, nevertheless, peculiarly human and strangely widespread. From what I understand, some heterosexual men occasionally and inexplicably indulge in homosexual activity, but if it is only out of lust, it can hardly be defined as bisexuality in its true sense. Human behaviour can indeed seem very strange and difficult to understand. On the other side of the coin, there must surely be some true homosexuals who wish that they weren't, especially since large sections of society still condemn homosexuality. It seems that gay activists have difficulty in grasping the possibility. Yet in my view, such men have a right to be covert if they so wish, and their privacy should not be invaded, unless their sexuality makes them too sad.

Homosexuality is but one of many biological aberrations, so there is no need to make a fuss about it. This, and other aberrations that originate before birth, are simply a fact of life. But because homosexuality, unlike other pre-natal disorders, is unusually widespread, it is tempting to wonder if it has evolved at some time during our biological history. However, it is not immediately apparent what advantage, if any, it has conferred or does confer on mankind – that is, how it might have been of benefit to our species. On the face of it, it appears to serve no beneficial biological purpose. It certainly does little towards population control, when we realize that the global village is vastly overpopulated. Such an interpretation, therefore, runs counter to the fundamental principle of evolution. Yet a homosexual man is not your ordinary run-of-the-mill male, so he has to be considered as being involuntarily but compulsively disordered. That homosexuality is uniquely human among mammals does not help us either, because there are many uniquely human disordered conditions and the fact that the homosexual condition is so common, no doubt arises from the fact that, thus far, it cannot be treated or detected prenatally. However, the complexity and fragility of our genetic make-up and of foetal hormones is such that any form of pre-natally based abnormality is to be expected as part of a wide spectrum of variation. There can, therefore, be no acceptable reason for homosexuality to be frowned upon any more than any other human condition. We accept people with Klinefelter's syndrome, Turner's syndrome, Down's syndrome and other syndromes, which arise from chromosomal deficiencies, which are also pre-natal and might affect sexual performance, so there is no reason why homosexuality, if recognized, should not be equally accepted. It seems perhaps that it is not tolerated by some people in society simply because it involves aberrant sexual behaviour rather than any other kind of behaviour. Moreover, the condition is not always manifest in facial or other physical features, although in some men it is, together with recognizable mannerisms. But if acts of homosexual sex are undertaken discreetly, as we expect 'normal' sexual activity to be, they can have no adverse effect on a society, because no one gets physically hurt. I predict that as more and more people come to recognize this, homophobia will ultimately die out altogether. The odious physical manifestations of it (an excuse for bullying?) are only expressed by barbarians.

The Anglican Church seems to have had something of a problem over the issue of homosexuality. This is understandable, because no one could, at this time, seriously expect a national church to condone sodomy. But whether or not one considers homosexuality as having evolved and thus to be simply representative of biological diversity, it has to be regarded, in the present state of our knowledge, as a clinical condition and not a simple choice of lifestyle. A desire on the part of a man to bond with another man is just not normal, because it is compulsive and nothing to do with choice. I believe that it would be helpful if this premise were to be taken on board by everyone, including theologians, and it seems reasonable that love between two individuals, whatever their sex may be, should have our blessing and that of the Church. By contrast, since a marriage ceremony concerns bonding and reproduction, there is a profound lack of logic in the advocacy of gay marriages, because such relationships only involve bonding.

The thought of anal copulation is distasteful to some people, but if it is undertaken as an act of love, is it really rational to condemn it? Such acts are not exclusively homosexual, and some homosexuals also dislike the idea. Thus, we come back to the importance of differentiating between sexual acts that stem from love and those that are undertaken purely out of lust. These are moral, legal and theological issues which are not really within the ambit of this text, so we must move on.

OTHER SEXUAL DISORDERS

Transvestite men are not always homosexual and since the activity involves no physical disorders nor hormonal ones, it must be construed as a purely psychological deviation from the norm. Transsexuals or intersexuals are different in that they have a permanent desire to change their sex and live life accordingly. These individuals often seek medicinal and surgical interventions in order to do so. (Female transsexuals may have abdominally situated, even if nonfunctional, testes, and in some cases they might even secrete testosterone.) Here, we are dealing in either sex with pre-natal developmental abnormality, be it physical or psychological. Accordingly, each case can only be properly assessed by experts with experience in this field.

No-one would deny that paedophilia is a disorder, but if you were a chimpanzee, you wouldn't worry about that either. Humans take a long time to mature, however, and since society properly feels the need of a collective obligation to protect its young from sexual predation, some form of firm restraint of paedophiles becomes urgent. In this context, I can only present some personal views.

In attempting to understand this horrible disorder, it seems to me to be important not to underplay the crucial role of imprinting on sexual development. If the first sexual experience of a human individual occurs pre-pubertly or neopubertly but with an adult, and it turns out to be pleasurable rather than, or as well as, repugnant, an association between childhood and pleasurable sex might be imprinted and the individual might fail to mature sexually. The experience could be frozen in the mind for life and be irreversible. This could apply to certain forms of homosexuality too, and if there is anything in the possibility, the age of consent at 16 comes perilously near to being too low. Can we really be sure that a peculiar sexual experience for a 16-year-old might not cause maturation damage? It depends on whether or not we consider 16-year-olds to have completed their sexual and social maturation. I doubt if they have, particularly most boys. Or could it also be that some pubertal children have a genetically based desire for sexual deviation anyway? The serious nature of paedophilia demands that more research on the possible origins and treatment of this presently incurable condition is urgently required.

Unlike paedophilia, in which victims are seriously hurt if not killed, homosexual activity between consenting adults does no harm to either party. It seems absolutely essential and logical, therefore, that compulsive paedophiles be not allowed to live in a community where they have easy access to young children. Most of these men use many wiles to satisfy their need, and monitoring is unlikely to be sufficient restraint.

I remember, as a 16-year-old boy, writing an essay on population control and in it, I advocated castrating all 'undesirable' males. My biology master wrote in the margin 'I think this is a bit extreme'! Of course, my youthful ardour needed to be cooled a little, but for paedophiles, mandatory chemical castration by using anti-androgens such as cyproterone acetate or more updated ones, might be considered as one possible approach to the problem. This is not as drastic as at first it seems, because it is often requested by the patients themselves. The difficulty is that if imprinting is involved, even this would not be a total answer.

Bestiality or zoophilia (copulating with a species other than your own) is neither typically primate nor mammalian behaviour and it is hard to accept as being normal. That is, of course, unless you are one of those cows who jumps on men as they cross a field! This sort of human behaviour reminds us again of the varying strength of sexual stimuli tested in the rabbit. Simply because it is alive and animated – it could perhaps be better a living sheep for some men, than a sex toy or a blow-up female doll! But surely, behaviour of this kind must be purely lustful and an act of desperation rather than of fundamental desire. Fortunately, evidence for and admissions of bestiality indicate that it is rare.

RAPE

Forced copulation occurs in species other than humans and has been recorded, for instance, in birds. But here again, neither species displays periods of oestrus. It has been reported in other primates such as macaque monkeys, when they are not in heat, so it could be a primate thing among mammals. Theoretically, human rape could result from

a spontaneous surge of testosterone, as with men who expose themselves in public. It would be akin to musth in elephants. They know
they shouldn't do it socially, but they do it just the same. Androgens
can lead men who are short on self-control to do things they really
shouldn't! Moreover, an excess of alcohol does little for self-control.

On the other hand, it is widely acknowledged that rape may be
an act of aggression as much as one of sexuality, and it is frequently
associated with psychological disturbance. I suspect, though, that with
a one-off rapist who otherwise appears to be normal, a rush of testosterone might start the process and aggression might follow ejaculation.
This would explain why some rapists kill their victims. For some
reason, a small number of male individuals become aggressive after
orgasm. This happens with some stallions, who will attack their hand-
ler after having covered a mare. Fortunately, it does not seem to occur
too often and most men are gentle and loving after ejaculation, if
perhaps a little exhausted (perhaps because they *are* exhausted!).

The problem of the proof of rape will never easily be resolved.
Because the human female has no periods of desire, how can a man be
absolutely sure whether she is receptive or not, especially if he does
not know her very well and may be drunk at the time? She may give a
signal, even accidentally, which gives the man a wrong impression. If
there are no witnesses, how can rape be proved or disproved? Sadly for
judges and juries, only they can decide. This lack of oestrus in human
females certainly causes problems and we might remember that when
other female mammals are out of season and thus have no periods of
desire, males also become sexually quiescent, but this never happens
in the human male, because neither oestrus nor seasonality exist.

AGE

It is established that after the age of 60, testosterone levels in men
decline, but there is no regular pattern as an outcome. To begin with,
the libido of individuals varies markedly, and since we have already
seen that penile erection has an imprint element to it (that is, it is not
totally governed by testosterone), a slight decrease in testosterone
levels may have little influence on erection in some men as they

age. Evidence is, however, that in older men, spontaneous erections normally become less frequent, weaker and less sustainable, even if the urge to copulate remains. Hence the popularity of Viagra and other such products. Incidentally, Viagra or Sildenafil, usually marketed as Sildenafil citrate, does not work on the nervous reflexes for erection (via the nervi erigentes), which normally cause dilation of the blood vessels of the penis, but rather acts locally and directly on the smooth muscle lining the helicine arteries of the penis, causing them to dilate and fill the cavernous sinuses with blood. However, it does not normally produce erection unless the man is sexually excited, or has reasonable levels of sensitizing circulating testosterone. So it seems as if the product simply augments normal – if weakened – processes for erection. But older men should be on guard! I quote from a cutting in the *Daily Express*, which someone sent me.

> German doctors say that three percent of men who die of heart attacks while having sex are in their late fifties and are making love to women 20 years younger. Seventy five percent of the women are mistresses!

Obviously, this serial monogamy of ours can be quite dangerous! But penile dysfunction in ageing men doubtless more commonly results from the impairment of general blood flow than to a decline in testosterone. This is why Viagra is mechanically so successful. But if testosterone levels fall too low, as they could in some ageing individuals, Viagra will not work either.

However, if endogenous circulating testosterone is not being used as much for purposes of sperm production or for expressions of libido, it could indeed become surplus to requirements in other areas. This is expressly apparent in the prostate, where the accumulated levels of testosterone prove too much for a failing organ and soon give rise to an enlarged prostate (prostatic hyperplasia). The condition is also quite common in older dogs in which the anatomical position of an enlarged prostate gives rise to constipation by causing stenosis (narrowing) of the rectum. In fact, the most common cause of chronic constipation in a dog is enlargement of the prostate. In contrast, the condition in man causes narrowing of the neck of the urinary bladder and interferes with micturition (urination). Also, on occasion, it can give rise to virtually chronic erection ('priapism') in man.

Formerly, prostatic hyperplasia in dogs was treated with female sex hormone (Stilboestrol), to counteract the accumulated level of testosterone in the area, but in man, more recently, a more imaginative treatment using Finesteride has come to be used. This drug inhibits the enzyme that converts testosterone into dihydrotestosterone (5-DHT), and since DHT is the active principle in the prostate rather than testosterone itself, this approach can be very effective. It presumably depends upon whether testosterone receptors also begin to fail with age.

EFFECTS OF MODERN TECHNOLOGY

The discovery and marketing of contraceptive pills have proved to be a major influence on changing human mating mores. This is because sexual gratification can be acquired without bonding and with no consequence, whereas before, copulation without barriers to the passage of sperms could only result in reproduction. Moreover, the application of such barriers undoubtedly interfered with spontaneity. The desire for unfettered coitus has been, and still is, a real problem in the spread of AIDS in Africa.

The contraceptive pill has been a boon to family planning, and sexual intercourse without kerfuffle or anxiety (through interruption) should theoretically help to secure a bond. The downside is that it has encouraged promiscuity, especially among young people and, paradoxically, may tempt older men to make their natural monogamy even more serial! This is not to mention the pill's significant role in the spread of sexually transmitted diseases (STDs). Because the pill is not yet widely used in Africa, it cannot be blamed for the devastating spread of HIV/AIDS on the continent, but it might have contributed to its spread in Europe and Asia, and could also be a factor in any increased incidence of teenage copulation.

Perhaps early on it might have been wise to have restricted the use of contraceptive pills to bonded couples. Although difficult to monitor, it could possibly have been done through legislation. But as often happens, governments and society generally were caught unawares by a sudden and tremendously rapid development and production of relatively safe contraceptive pills. It all happened so quickly that, in Britain, an increased incidence of STDs, which is not surprising, was apparently not predicted when contraceptive pills became freely available. But something has gone wrong, because this has also been accompanied by an increase in the rate of unwanted or unplanned pregnancies in Britain. This could, of course, reflect a simple ignorance on the part of many young people concerning the basic principles of reproductive physiology and the importance of self-restraint in all sorts of different aspects of life. So much for sex education! However, the difficulty with any form of legislation is that somebody will usually find a way around it, so it is extremely difficult to draft adequately.

Until the advent of 'the pill', we operated a society that was essentially a latter-day form of hunter-gatherer society. Women reared the children and men were the providers (in other words, the hunters). Safe and simple contraception has changed all that and has encouraged women to move into the metaphorical hunting field. Not that they might not have done that anyway. Working spinsters might remember, however, that if we are typically primate, it is the bonded females who have reproduced that are the dominant ones in our communities!

I fear that several facets of political correctness conflict with many of our biologically based instincts. But we have to remember that biology does not deal with individual rights, it deals with the survival of species. Unfortunately, when 'causes' come through the door, rationality often flies out of the window. Take having babies at the best time (20–35 years of age for a woman) as an example. It has lost its appeal for many women, now that it need not happen. Yet from a physiological, genetic, demographic and economic standpoint, it will be to our advantage in the West if we were to try and reverse this trend, or at least slow it down. It would take time, but in the long run, the portents are that it will be better for us if we indulge in more widespread mating for purposes of reproduction, rather than it being used purely as an exercise in pleasure.

Young people are crucial to economic progress, if not economic survival, so families could beneficially be encouraged to reproduce more freely, in spite of the mother being in the workforce. Thus, we, in the Western world especially, might consider having slightly larger families. The only other approach to the problem is to import young people into the community through immigration, but this can only put off the evil hour and leave yet another generation to find a solution, which could become increasingly elusive. The achievement of balance between the options is the lot of politicians, so we shall see what they do. Whatever we do globally, it must also be balanced with careful control of reproduction in areas of the world where there is serious over-population and food shortage already. Here we have an even more intractable problem, because if we try to solve over-population by migration, we are sure to run into massive demographic changes and into difficulties that might arise from an indiscriminate mixing of cultures. It is clear, therefore, that in our ever-increasing ability to control our environment, the Western world has created long-term biological problems for itself, because we are tending to forget that mating is primarily part of the process of reproduction, and in the world of living things, large numbers are influential and usually result in dominance. We want every human being to have a reasonable standard of living, but evidence shows us that economic growth invariably leads to a reduction in the size of families.

Next, in the chronology of human technological advance, came in vitro fertilization (IVF), eventually including the freezing of sperms and embryos. A blessing for the infertile woman, IVF, to my mind, is almost as significant a scientific achievement as the discovery of the structure of DNA, and its importance has been grossly under-estimated until recently. But IVF, too, has brought with it manifold problems or potential problems, the main one being that it allows human reproduction to be achieved without mating. Thus, we are now witnessing a progressive schism between copulation and reproduction. Combine this trend with the contraceptive pill and we have a society in which bonding is no longer a prerequisite of our reproduction. In the longer term, this must exert a profound effect on our culture, if it has not already done so. Right before our eyes now, we see copulation as being purely part of the pursuit of pleasure for some people, at the same time as reproduction by means of various and sometimes impersonal technologies is increasing in popularity. It seems to be the case that this is because reproduction can now be achieved without the actual presence of a male. But if particular semen donors become too popular, it can only have an adverse effect on the genetic pool by reducing a more random mix of genes. Moreover, with semen donors generally, without strict rules, we can run into consanguinity problems.

Instant hedonism is currently all the rage, and we rejoice in pleasure without unfortunate consequences. But when it comes to sexual pleasure, it cannot, in the long run, but wreak societal change. With a rapidly developing technology such as IVF, we must be ready and prepared for this. The passionate voice of Eros now very much dominates the gentler touch of Demeter.

More recent developments include use of the technique of intra-cytoplasmic sperm injection (ICSI). With this method, an individual sperm head can be injected directly into an egg. It is a wondrous thought that a man who can produce no sperms at all in his ejaculates, and if this be only due to obstruction in the tubular system, can still father a child. This is because sperms can now be taken directly from the testis and used in IVF, or a single sperm can be selected for ICSI. The problem is that testicular sperms, as we have seen, are theoretically immature, and if this immaturity involves the condition of DNA as well as other features of the sperm, then who knows what long-term effects there might be? This still holds true even if babies produced by such a method seem fine and dandy when they are born.

Only time will tell, but in our enthusiasm, by using techniques of this kind (unlike IVF and ICSI using reproductively healthy males), we are getting near to ethically dangerous ground and can only hope that nothing goes wrong. Fortunately, the use of testicular sperms in IVF or ICSI is rare. However, we should not ignore the possibility that all modern in vitro technologies, including IVF, might lead us to perpetuate genetically based infertility. How and when should we try to balance the needs of sub-fertile individuals and their happiness with the long-term fitness of populations? It is an important question, because if it is not addressed, some people may be tempted into believing that males are becoming increasingly redundant in terms of human reproduction. Sperms can now be produced *in vitro* from spermatogonia, for example, and several females can be inseminated using a single donor.

A belief, widely – and to my mind incautiously – held by many people, including some scientists, is that on top of this, human sperm count is on the decline worldwide. This is the sort of story the media thoroughly enjoy. As a result, it has become a popular notion among the public at large. It may well be that sperm count in some parts of the globe is decreasing, apparently more so in Denmark than Finland, but evidence that it is a serious worldwide trend is certainly not established. Even if it were, its significance would be difficult to assess. We have no idea what the sperm count was among the ancient Greeks or of men of the Roman Empire. It is distinctly possible that over several centuries human sperm count has fluctuated spontaneously or for some special reason. With equal spontaneity, it could recover.

We should not forget that a slight reduction in male fertility does not mean total infertility. In areas such as Europe, reduced rates of reproduction are much more likely to be the outcome of choice, dangerous though that may be. In a Wonderland world, however, the Alices might come to demand a sperm count before they are prepared to bond. For those who like to claim that men are the inferior sex, a downturn in sperm count is all grist to the mill. What a delight for them it could be, if it could also be demonstrated that men are, in this modern world, getting more inferior by the minute!

Adding fuel to this fire is the recognition that the Y chromosome is a degenerate form of the X chromosome. A lunatic fringe might take this as another sign of the disappearing male. However, early on in embryological development, we have seen that one of the X chromosomes in females is inactivated, so adults, be they male or female, end up with the same complement of active X. But since the Y chromosome simply endows the individual with maleness, it doesn't matter whether it is a degenerate chromosome or not.

Moreover, the Y chromosome is versatile and has very good powers of regeneration. Less than 100 per cent of a Y chromosome is Y anyway. It has a complicated evolutionary history, so I don't think denigrators or potential denigrators of males have, on this score, much of a leg to stand on. Males are here to stay, however much some people might like to wish them away. And they are not just playthings either, because they are essential in maintaining the vitality of the population.

The most recent developments in reproductive technology include animal cloning, the freezing of surplus human embryos and the use of single embryonic and adult cells in stem cell research, as well as in pre-implantation diagnosis of disease. Prospects for these techniques are immense, and it must be acknowledged that, whatever prejudices and beliefs one may have, these new advances have provided knowledge that has offered more gladness and joy than sorrow to those people in need of assisted reproduction. It is surely unacceptable to scare people unnecessarily with talk of 'designer babies'. They may ultimately happen to the extent of the issue being important, but not for a long time yet.

Nevertheless, science, even if some of its potential developments are carefully monitored, must be permitted to proceed essentially unhindered. Even if freezing of eggs and embryos becomes a vogue in the future, and people, through Wellsian indifference to current moral values, are able eventually to choose their sperms, their eggs and their embryos (even over the counter) and whatever new scientific advances, such as cloning, may bring to the field of human and animal reproduction, some things are unlikely ever to change. For instance, I think we can be assured that, for an awfully long time yet, males, even human males, will still be mating and proffering their genes, just as they have always done.

© iStockphoto/Clint Scholz.

FURTHER READING

Abbott, D.H. & Hearn, J.P. (1978) Physical, hormonal and behavioural aspects of sexual development in the marmoset monkey (*Callithrix jachus jachus*). *Journal of Reproduction and Fertility*. **63**, 335–345.

Baldwin, J.D. (1970) Reproductive synchronization in squirrel monkeys (*Saimiri*). *Primates*. **11**, 317–326.

Dixson, H.F. & Gardner, A.F. (1983) Diurnal variation in plasma testosterone in a male nocturnal primate, the owl monkey (*Aotus trivirgatus*). *Journal of Reproduction and Fertility*. **62**, 83–86.

Dukelow, W.R. (1983) The squirrel monkey (*Saimiri squireus*). In: *Reproduction in New World Primates*. Ed.: J.P. Hearn. MTP Press Ltd, Lancaster, Boston and The Hague.

Epple, G. (1975) The behaviour of marmoset monkeys (Callithricidae). In: *Primate Behaviour*. Ed.: I.R. Rosenblum. Vol. **4**. Academic Press, New York.

Hearn, J.P. & Lunn, S.F. (1975) The reproductive biology of the marmoset monkey: *Callithrix jacchus*. *Laboratory Animal Handbooks*. **5**, 191–202.

Hinde, R.A. (1971) Development of social behaviour. In: *Development of Non Human Primates*. Vol. **3**. Eds: A.M. Schreier & F. Stollnitz. Academic Press, New York.

Hinde, R.A. & Spencer-Booth, Y. (1968) The study of mother–infant interaction in captive group-living rhesus monkeys. *Proceedings of the Royal Society. London. Series B*. **169**, 177–201.

Johnson, A.M., Mercer, C.H., Erens, B., Copas, A.J., McManus, S., Wellings, K., Fenton, K.A., Korovessis, C., Macdowell, W., Nanchahal, K., Purdon, S. &

Field, J. (2001) Sexual behaviour in Britain: partnerships, practices and HIV risk behaviours. *Lancet.* **358**, 1835–1842.

Jones, S. (2003) *Y: The Descent of Men.* Abacus, London.

Jost, A. (1946) Sur la differentiation sexuelle de l'embryon de lapin, remarke au sujet de certaines operations chirurgicalles sur l'embryon. *Compte rendue de la Société biologie.* **140**, 460–463 (if the reader is unable to read French, they might care to consult Jost (1965) instead).

Jost, A. (1965) Gonadal hormones in the sex differentiation of the mammalian testis. In: *Organogenesis.* Eds: R.L. de Hahn & R. Ursprung. Holt, Rinehart & Winston, New York.

Latta, J., Hopf, S. & Ploog, D. (1967) Observations on mating behavior and sexual play in the squirrel monkey (*Saimiri squireus*). *Primates.* **8**, 229–246.

Leakey, R. & Lewin, R. (1992) *Origins Reconsidered.* Little Brown & Co (UK) Ltd, London.

Llewellyn-Jones, D. (1984) *EveryMan.* Oxford University Press, Oxford.

Lyndon, N. (1992) *No More Sex Wars.* Sinclair-Stevenson, London.

Michael, R.P. (1971) Determinants of primate reproductive behaviour. In: *WHO Symposium on the Use of Non-human Primates in Research on Human Reproduction.* Sukhumi, USSR.

Michael, R.P. & Keverne, R.B. (1968) Pheromones in the communication of sexual status in primates. *Nature. London.* **218**, 746–749.

Morris, D. (1977) *Manwatching.* Jonathan Cape, London.

Morris, D. (1999) *The Naked Ape.* Dell, New York.

Nagle, C.A., Denari, J.H., Riarke, A., Quiroga, S., Zarate, R., Germino, N.I., Merlo, A. & Rosner, J.M. (1980) Endocrine and morphological aspects of the menstrual cycle in the cebus monkey (*Cebus apella*). In: *Non Human Primates for Study of Human Reproduction.* Karger, Basel.

Napier, J.R. & Napier, P.H. (1967) *A Handbook of Living Primates.* Academic Press, London and New York.

Parkes, A.S. (1963) *Sex, Science and Society.* Oriel Press Ltd, Newcastle-upon-Tyne.

Pearsall Smith, L. (1945) *All Trivia.* Constable & Co. Ltd, London.

Potts, M. & Short, R. (1999) *Ever Since Adam and Eve.* Cambridge University Press, Cambridge.

Short, R.V. (1980) The great apes of Africa. *Journal of Reproduction and Fertility: Supplement.* **28**, 3–11.

Van den Berghe, P.L. (1979) *Human Family Systems.* Elsevier, New York.

Appendix

Mammals are referred to in the text by their popular names only, so the list below provides, in sequence, the order, family and genus or species of each animal mentioned.

Common name	Order	Family	Species
Aardvark	Tubulidentata	Orycteropodidae	*Oryctopus afer*
African jumping hare or springhaas	Rodentia	Pedetidae	*Pedetes capensis*
Alpaca	Artiodactyla	Camelidae	*Lama pacos*
Anteaters	Edentata	Myrmecophagidae	*Myrmecophaga tridactyla* (giant anteater); *Cyclopes didactylus* (dwarf anteater)
Antelopes			
Blackbuck	Artiodactyla	Bovidae	*Antelope cervicapra*
Dik-dik	Artiodactyla	Bovidae (Antilopinae)	*Madoqua* (various species) and *Dorcatragus*
Eland (very large with spiral horns)	Artiodactyla	Bovidae	*Taurotragus oryx*
Impala	Artiodactyla	Bovidae	*Aepyceros melampus*
Kudu (fairly large with spiral horns)	Artiodactyla	Bovidae	*Tragelaphus septiceros*
Springbok	Artiodactyla	Bovidae	*Antidorcas marsupialis*
Waterbuck	Artiodactyla	Bovidae	*Kobus ellipsiprymnus*
Includes also the gazelles, such as *Gazella thomsoni* (Thomson's gazelle) and *Gazella granti* (Grant's gazelle), Blesbok and Lechwe			
Armadillo (nine-banded)	Edentata	Dasypodidae	*Dasypus novemcinctus*
Badger	Carnivora	Mustelidae (Melinae)	*Meles meles*
Bats			
Common bat	Chiroptera	Vespertilionidae	*Myotis* (many species)
Long-eared bat	Chiroptera	Phyllostomatidae	*Micronicteris* (several species)
Pipistrelles	Chipotera	Vespertilionidae	*Pipistrellus* (many species)

Common name	Order	Family	Species
Beaver	Rodentia	Castoridae	*Castor fiber*
Bears			
Brown bear	Carnivora	Ursidae	*Ursus arctos*
Polar bear	Carnivora	Ursidae	*Thalarctus maritimus*
Bison	Artiodactyla	Bovidae	*Bison bison*
Blesbok	Artiodactyla	Bovidae	*Damaliscus dorcas*
Buffalo (water)	Artiodactyla	Bovidae	*Syncerus caffer*
Camel	Artiodactyla	Camelidae	*Camelus bacterianus*; *Camelus dromadarius*
Cats			
Domestic cat	Carnivora	Felidae	*Felis catus*
Cheetah	Carnivora	Felidae	*Acinomyx jubata*
Jaguar	Carnivora	Felidae	*Panthera onca*
Leopard	Carnivora	Felidae	*Panthera pardus*; *Panthera uncia* (snow leopard)
Lion	Carnivora	Felidae	*Panthera leo*
Lynx	Carnivora	Felidae	*Felis lynx*
Tiger	Carnivora	Felidae	*Panthera tigris*
Chamois	Artiodactyla	Bovidae	*Rudicapra rudicapra*
Chevrotain	Artiodactyla	Tragulidae	*Tragulus meminna*
Chipmunk	Rodentia	Marmotidae	*Tamias* (several species)
Civet	Carnivora	Viverridae	*Viverra civetta*
Cow (ox)	Artiodactyla	Bovidae	*Bos taurus*
Coyote	Carnivora	Canidae	*Canis latrans*
Deer	Artiodactyla (Ruminantia)	Cervidae	Many species, but reference in text is to red deer, *Cervus elaphus*; and fallow deer, *Cervus dama*
Dog	Carnivora	Canidae	*Canis familiaris*

Dolphin	Cetacea	Delphinidae	*Tursiops truncatus* (bottle-nosed dolphin) (several species and other genera) (reference in text is to *Tursiops truncatus*)
Dormouse	Rodentia	Gliridae	*Muscardius avellanarius* (several genera)
Duck-billed platypus	Monotremata	Ornithorhynchidae	*Ornithorhyncus anatinus*
Dugong	Sirenia	Dugongidae	*Dugong dugon*
Duiker	Artiodactyla	Bovidae	*Cephalophus adersi*
Echidna	Monotremata	Tachyglossidae	*Tachyglossus aculiatus*
Elephant	Proboscidea	Elephantidae	*Loxodonta Africana* (loxodonta) *Elephas elephas* (maximus)
Ferret	Carnivora	Mustelidae	*Mustela furo*
Field mouse	Rodentia	Muridae	*Apodemus* (several species); *Micromys minutus* (harvest mouse)
Fox	Carnivora	Canidae	*Vulpes vulpes*
Giraffe	Artiodactyla	Cervidae	*Giraffa camelopardalis*
Gnu	Artiodactyla	Bovidae	*Taurotragus* (two species)
Goat	Artiodactyla	Bovidae	*Capra hircus*
Golden mole	Insectivora	Chrysochloridae	*Chrysochloris* (several species)
Gopher	Rodentia	Geomyidae	e.g. Genus *Geomys*, but several genera and species
Guinea pig (Cavy)	Rodentia	Cavidae	*Cavia porcellus*
Hamster	Rodentia	Cricetidae	*Cricetus auratus*
Hare	Lagomorpha	Leporidae	*Lepus arcticus* (Arctic hare); *Lepus europaeus* (brown (European or common) hare); *Lepus americanus* (snowshoe hare)
Hartebeest	Artiodactyla	Bovidae	*Alcephalus buselaphus*
Hedgehog	Insectivora	Erinacidae	*Erinaceus europaeus*

Hippopotamus	Artiodactyla	Hippopotamidae	*Hippopotamus amphibius*
Horse	Perissodactyla	Equidae	*Equus caballus*
Hyena	Carnivora	Hyaenidae	*Crocuta crocuta* (spotted hyena); *Hyaena hyaena* (striped hyena)
Hyrax (rock)	Hyracoidea	Procavidae	*Procavia capensis* (South Africa, where it is known as the 'Dassie'); *Procavia capensis* (throughout the Rift Valley in Kenya and the Middle East). Also *Heterohyrax brucei* in Kenya, *Procavia habessinica* and then *Procavia capensis syriacus* on the Golan Heights
Hyrax (tree)	Hyracoidea	Procavidae	*Dendrohyrax arboreus*
Jackal	Carnivora	Canidae	*Canis mesomelas* (black-backed (or grey-backed) jackal and a few other species)
Kangaroo (red)	Marsupiala	Macropodidae	*Macropus rufus*
Kinkajou	Carnivora	Procyonidae	*Potus flavus*
Kudu	Artiodactyla	Bovidae	*Tragelaphus strepsiceros*
Llama	Artiodactyla	Camelidae	*Lama glama*
Lechwe	Artiodactyla	Bovidae	*Kobus leche*
Lemur	Primates (prosimian)	Lemuridae	Several species. *Lemur catta* (ring-tailed lemur)
Loris	Primates (prosimian)	Lorisidae	*Nycticebus coucang* (Slender loris (Asia); slow loris (Asia)); *Perodycticus potto* (Potto (e.g. Bosman's, Africa))
Man	Primates	Hominidae	*Homo sapiens* (also mentioned, early man, *Homo erectus*)
Manatee	Sirenia	Trichechidae	*Trichechus monatus*

Marmot	Rodentia	Sciuridae	*Marmota marmota* (a few different species)
Marsupial mouse	Marsupialia	Dasyuridae	*Antechinus* (several species)
Mink	Carnivora	Mustelidae	*Mustela lutreola* (European mink); *Mustela vison* (American mink)
Mole rat (Cape)	Rodentia	Bathyergidae	*Bathyergus suillus*
Mongoose	Carnivora	Viverridae	*Herpestes auropunctatus*
Monkey	**Primates**		
New World monkeys (– anthropoids)			
Cebus monkey or capuchin	Primates	Cebidae	E.g. *Cebus capucinus* (white-throated capuchin)
Colobus monkey	Primates	Cebidae	*Colobus polykomos* (black-and-white colobus)
Howler monkey	Primates	Cebidae	A few species, e.g. *Alouatta seniculus* (red howler monkey)
Marmoset	Primates	Callithricidae (also Tamarin)	*Callithrix jacchus*; *Leontideus rosalia* (lion tamarin)
Owl (night) monkey or dourocoulis	Primates	Cebidae	*Aotus trivirgatus*
Spider monkey	Primates	Cebidae	*Atteles geoffroyi* (a number of sub-species) and the genus *Agothrix* (woolly monkeys)
Squirrel monkey	Primates	Cebidae	*Saimiri sciureus*
Titi monkey	Primates	Cebidae	*Callicebus cupreus* (several species)
Old World monkeys (– anthropoids)			
Baboon	Primates	Cercopithecidae	*Papio hamadryas* (Africa and the Middle East). Also, several African species, including the Gelada baboon, *Theropithecus gelada*
Chimpanzee	Primates	Pongidae (apes)	*Pan troglodytes*
Gibbon	Primates	Pongidae	*Hylobates* (several species)

Gorilla	Primates	Pongidae	*Gorilla gorilla* (anthropoid ape)
Macaque (rhesus) monkey (anthrapoid)	Primates	Cercopithecidae	*Macaca mulatta*
Orang utan (anthropoid ape)	Primates	Pongidae	*Pongo pygmaeus*
Mountain goat	Artiodactyla	Bovidae	*Oreamnos americanus*
Mouse	Rodentia	Muridae	*Mus musculus* (house mouse, but many species, in this genus). Also, *Apodemus* (field mouse – a few species)
Muskrat	Rodentia	Cebidae	*Ondotra zibotheca*
Opossum	Marsupialia	Didelphidae	*Didelphis marsupialis* (common possum of America)
Opossum	Marsupialia	Phalangeridae	*Trichosurus vulpecula* (brush-tailed possum of Australia)
Opossum	Marsupialia	Phalangeridae	*Tarsipes rostratus* (Australian honey possum)
Otter	Carnivora	Mustelidae	*Lutra lutra* (but several species)
Panda	Carnivora	Procyonidae	*Ailuropida melanoleuca* (giant panda); *Ailurus fulgens* (red panda)
Pangolin	Pholidota	Manidae	*Manis* (a few different species)
Pig	Artiodactyla	Suidae	*Sus scrofa* (domestic pig and wild boar)
Polecat	Carnivora	Mustelidae	*Mustelus putorius*
Porcupine			
Old World	Rodentia	Hystricidae	E.g. *Hystrix* (several species)
New World		Erethizonditidae	E.g. *Erethizon dorsatum*

Common name	Order	Family	Genus/species and notes
Porpoise	Pinnipedia	Delphinidae	*Phocaena phocoena*
Potto	See Loris		
Raccoon	Carnivora	Procyonidae	*Procyon lotor* (but several species)
Rabbit	Lagomorpha	Leporidae	*Oryctolagus* (*Lepus*) *cuniculus*
Rat	Rodentia	Muridae	*Rattus norvegicus*; also *Rattus rattus*
Rhinocerus	Perissodactyla	Rhinocerotidae	Two-horned African *Dicerus sinus* or *Ceratotherium simum* (white rhino); *Dicerus bicornis* (black rhino); *Diderocerus* or *Dicerorhinus sumatrensis* (two-horned Sumatran); *Rhinoceros* (*unicornius* and *sondaicus*) single-horned Asian (Indian and Javan, respectively)
Seal	Pinnipedia	Phocidae	*Phoca vitulina* (common seal); *Mirounga leonina* (southern elephant seal)
Sealion	Pinnipedia	Otariidae	*Zalophus californianus* (Californian sealion) *Neophoca cinerea* (Australian sealion) and other genera and species
Sheep	Artiodactyla	Bovidae	*Ovis aries*
Shrew	Insectivora	Macroscelididae	Many species and genera but *Elephantulus rozeti* (elephant shrew); *Petrodromus sultan* (East coast (African forest) shrew); *Rhynchocyon chrysopagus* (yellow-backed elephant shrew)
Skunk	Carnivora	Mustelidae	*Mephitis mephitis* (several genera and species)
Sloth	Edentata	Bradypodidae	A few different species. *Bradypus tridactyla* (three-toed sloth); *Choelepus didactilus* (two-toed sloth)
Solenodon	Insectivora	Solenodontidae	*Atopigale cubana, Solenodon paradoxus*

Squirrel	Rodentia	Sciuridae	*Sciurus vulgaris* (red squirrel); *Sciurus carolinensis* (grey squirrel); but also, *Glaucomys volaris* (North American flying squirrel – several genera of flying squirrels) and *Citellus ticemlineatus* (ground squirrel – many species and other genera)
Stoat	Carnivora	Mustelidae	*Mustela erminea*
Tarsier	Primates (prosimian)	Tupaiidae	*Tarsius bancanus*
Tenrec	Insectivora	Tenrecidae or Centetidae	*Microgale* (the most widespread genus in Madagascar)
Tree shrew	Primates (prosimian)	Tupaiidae	*Tupaia glis* (common tree shrew), several species
Vole	Rodentia	Cricetidae	E.g. *Microtus* (field vole), many species and different genera, including *Clethrionomys* (bank vole) or *Evotomys* (prairie vole)
Walrus	Pinnipedia	Odobenidae	*Odobenus rosmarus*
Warthog	Artiodactyla	Suidae	*Phacochoerus aethiopicus*
Weasel	Carnivora	Mustelidae	*Mustela nivalis*
Whale	Cetacea	Several families	Many species (see Bryden, Marsh and Shaughnessy (1998), referenced on p. 113 of this volume). Pilot whale cited in the present text (*Globiocephala melaena*)
Wolf	Carnivora	Canidae	*Canis lupus*

The above information has been widely sourced, but Morris, D. (1965) *The Mammals*. Hodder & Stoughton, London (in association with the Zoological Society of London) has proved to be particularly helpful in cross-checking.

Index

Printed in the United States
by Bookmasters Publisher Services

Printed in the United States
by Baker & Taylor Publisher Services